PSP™
HACKS, MODS, AND EXPANSIONS

PSP™
HACKS, MODS, AND EXPANSIONS

DAVE PROCHNOW

McGraw-Hill

NEW YORK | CHICAGO | SAN FRANCISCO | LISBON
LONDON | MADRID | MEXICO CITY | MILAN | NEW DELHI
SAN JUAN | SEOUL | SINGAPORE | SYDNEY | TORONTO

The McGraw·Hill Companies

Copyright © 2006 by The McGraw-Hill Companies, Inc. All rights reserved. Printed in the United States of America. Except as permitted under the United States Copyright Act of 1976, no part of this publication may be reproduced or distributed in any form or by any means, or stored in a data base or retrieval system, without the prior written permission of the publisher.

1 2 3 4 5 6 7 8 9 0   DOC/DOC   0 1 0 9 8 7 6 5

ISBN 0-07-146908-7

The sponsoring editor for this book was Judy Bass, the editing supervisor was Stephen M. Smith, and the production supervisor was Pamela A. Pelton. It was set in ITC Officina Serif by Cindy LaBreacht. The art director for the cover was Anthony Landi.

Printed and bound by RR Donnelley.

McGraw-Hill books are available at special quantity discounts to use as premiums and sales promotions, or for use in corporate training programs. For more information, please write to the Director of Special Sales, McGraw-Hill Professional, Two Penn Plaza, New York, NY 10121-2298. Or contact your local bookstore.

 This book is printed on recycled, acid-free paper containing a minimum of 50% recycled, de-inked fiber.

Information contained in this work has been obtained by The McGraw-Hill Companies, Inc. ("McGraw-Hill") from sources believed to be reliable. However, neither McGraw-Hill nor its authors guarantee the accuracy or completeness of any information published herein, and neither McGraw-Hill nor its authors shall be responsible for any errors, omissions, or damages arising out of use of this information. This work is published with the understanding that McGraw-Hill and its authors are supplying information but are not attempting to render engineering or other professional services. If such services are required, the assistance of an appropriate professional should be sought.

CONTENTS

PREFACE vii

ACKNOWLEDGMENTS xiii

INTRODUCTION The Dawn of the Digital Lifestyle 1

CHAPTER 1 Yeah, But How Do I...? 17

CHAPTER 2 The Hack Heard 'round the World 51

CHAPTER 3 SAW: The PSP Dissected 57

CHAPTER 4 Shtick Talk 73

CHAPTER 5 UMDware 87

CHAPTER 6 The Modder of All PSPs 105

CHAPTER 7 Get Yer Game Face On 113

CHAPTER 8 Don't Forget the Sunscreen 125

CHAPTER 9 Get a Wi-Fi of This 145

CHAPTER 10 USB OTG—Let's Go 157

CHAPTER 11 Status Quo Vadis? 173

CHAPTER 12 PSP P.S. 195

EPILOGUE 197

APPENDIX A PS2Pdp 199

APPENDIX B Ten Games Ya Gotta Play 203

APPENDIX C Ten Movies Ya Gotta Watch 209

APPENDIX D Things to Know, Places to Go 215

INDEX 219

PREFACE

Not too long ago, around 1620 to be precise, the term "hacker" was used to denote a person who was unskilled or inexperienced at a particular activity. For example, you were a golf *hacker*; or, I was a painting *hacker*. Now fast forward approximately 350 years, and the term "hacker" had undergone a radical facelift.

Around 1970, rather than being unskilled or inexperienced at an activity, a hacker was considered to be a highly skilled, clever programmer or a technology expert. What? You thought a hacker was a criminal, a ne'er-do-well, a malcontent who thrives on crippling computer systems? Wrong-o, bucko. Let's set the record straight, once and for all.

As coined in the early 1970s, a hacker was a genius, a gifted individual who was the master of a software or hardware domain over which digital dominion was exercised. Ah, the good ol' days. Unfortunately, in the 1980s, a different definition for hacker came into mainstream usage. This metamorphosis didn't have to happen, however.

As far as hackers are concerned, the 1980s started in a heroic manner with the publication of Steven Levy's book *Hackers: Heroes of the Computer Revolution*. In 1989, however, the hyperanimated Clifford Stoll (author of *The Cuckoo's Egg*) penned a different take on hackers—as computer criminals. Soon, public awareness and media propagation embedded the hacker nee criminal concept into colloquial English. And, the negative connotation for hacker stuck.

Now, if you ask a hacker what to call a computer criminal, the most common reply would be a "cracker." According to hackers, a cracker is a thief, subversive, or criminal who is determined to break, cripple, or vandalize a computer network or system. Additionally, the term "black hat" might also be applied to this form of computer criminal. In this case, the appellation is derived from a negative nod to a popular flavor of Linux (i.e., Red Hat). And Linux is *the* most popular weapon in the arsenal of the hacker.

My Life in the Video Game

In the mid-1980s, I worked on several video game projects. These projects, *Flight Simulator*, *JET*, *GUNSHIP*, and *Bulls & Bears,* were each well-known games. My involvement was in the preparation of ancillary popular nonfiction books that would support these games. For example, I wrote a book (*Flight Simulator and Flight Simulator II: 82 Challenging New Adventures*, TAB Books, 1987) that was published independently from Microsoft with the sole intent of exploring and exploiting this flight simulator game. This book, as well as all of my other similar titles, was an unofficial work that involved a great deal of research and experimentation to learn and then write about all facets of game play—both documented and, especially, undocumented features.

During this early video game documentation work, I learned that the user's manual was largely misinformation and that most video game programmers embedded Easter eggs (hidden secret access points) and cheat codes within every product. For example, one of the all-time great cheat codes in *JET* was the ability to shoot down lots of enemy fighters just by flying in an "unconventional" manner (e.g., by circling the aircraft carrier, firing your missiles in a random sequence, and recovering on the carrier for immediate rearming and refueling—scores of 90+ "kills" could be regularly achieved with this technique). Quickly, Easter eggs and cheat codes became mainstays in most video games. Even operating systems (e.g., Mac OS System 7.5 included a hidden copy of *Breakout*) began to sprout Easter eggs for piquing the interest of the savvy computer user.

Gaining access to this type of hidden information can be a daunting task. You aren't going to just happen on an arcane keystroke sequence that, for example, will give you unlimited game strength. Yes, the "one thousand monkeys typing on one thousand typewriters for an infinite amount of time could write the complete works of Shakespeare" theory is possible, but a far more reliable and time-efficient method for unearthing hidden game traits is to hack.

An Interview with a Sony PlayStation® Portable Hacker

If it's a game hacker who you're trying to find, then there's no better place to look than the Electronic Entertainment Expo (E3). Hackers, hucksters, hipsters, and a fairly saucy number of hussies creep through the expo's halls and aisles in search of new video games and gaming hardware. This is the ideal lair for finding hackers and crackers. And this is where I met $onny7.

This hacker, like almost every one who I've ever met, uses a digital nom de plume in lieu of a more proper legal name. Also like the typical hacker, the only background or other personal information that you can glean from this individual is a list of hacks. Forget home addresses, educational achievements, and employment records. The only thing that a hacker will do is talk tech. In this case, $onny7 has been using, playing, and hacking on the Sony PlayStation Portable (PSP™) since its 2004 introduction in Japan.

DAVE PROCHNOW: There appears to be a pervasive negative consensus about the portrayal of hackers in the media. Are you a hacker or a cracker?

$ONNY7: Yes.

DP: That's a rather flip answer, isn't it? Doesn't such a glib response trivialize the question and make the media's incorrect portrayal seem more valid to the public?

$ONNY7: Yeah...so. I'm a hacker when I hack and a cracker when I crack.

DP: Give me an example of a crack.

$ONNY7: When I see some n00b leaving the back door open, I'll go inside and tell 'em to close the stupid door. That type of crack.

DP: N00b?

$ONNY7: Yeah, a newbie or beginner.

DP: So your cracks are noble services rendered to inexperienced computer users?

$ONNY7: Jeez, no; it's more like an act of civil disobedience that doesn't do anybody any harm, but serves as a wake-up call that you aren't as safe as you think you are.

FIGURE P-1 On March 24, 2005, the Sony PlayStation Portable took the video game industry by storm.

DP: You're the "white hat" cracker?

$ONNY7: LOL.

DP: So if that's your idea of a good crack, give me an idea of a good hack.

$ONNY7: Busting the screws off.

DP: Busting the screws off?

$ONNY7: Yeah, opening up a device, learning how it works, and making it work for you. Listen, you bought the product, right? You own it, right? Actually, wrong. Most consumer products like an Apple iPod are closed systems. You buy it, operate it the way the manufacturer tells you to operate it, and live within a rigid set of restrictions imposed by the manufacturer. That's just not right. You bought it, you own it, and you should be able to use it. My hacks let you take back your consumer electronics product from manufacturer-imposed limitations.

DP: Give me an example of a consumer electronics product that you've busted the screws off of.

$ONNY7: There's a couple of good ones right now, but the best one is the PSP.

DP: The Sony PSP?

$ONNY7: Yeah, here you've got the perfect multimedia gaming system, but you can't even make your own video movies display at H.264. Why not? It's your PSP. I'm working on hacks that allow users to gain control over their PSPs. Take back control of your PSP. Hack it.

DP: Do you announce your hacks online or participate in any Web forums?

$ONNY7: No. I work for me.

DP: So, how can others learn how to take back control of their PSPs?

$ONNY7: Hack it. EOF.

Dave Prochnow

ACKNOWLEDGMENTS

This is the best book ever written about using, hacking, modding, and expanding the Sony PlayStation Portable. It might not have turned out that way if Amelia, Anthony, Katherine, Kathy, Judy, and Penelope hadn't pulled out all of the stops and made significant, timely, and valuable contributions toward understanding, exposing, and recording everything about the PSP.

Likewise, without the support provided by Agetec, Aiptek, Belkin, bhv Software, Boxwave, Cowon America, Gamer Graffix, Geneon Entertainment, Image Entertainment, Junxion, Logic3, Logitech, Norêve, Otterbox, Pelican, SanDisk, SpectraVideo, track7games, and Vaja this book would have been nearly impossible to write.

Thank you one and all; you may now return to playing, listening to, and watching your PSP.

ABOUT THE AUTHOR

Dave Prochnow is an award-winning professional writer, editor, and contributor to numerous technical publications, including *MacAddict*, *Link-Up*, and *digitalFOTO*. He is the author of 25 nonfiction books for Addison-Wesley, F&W Publications, McGraw-Hill, and TAB Books, including the bestselling *The Official Robosapien Hacker's Guide* and *Experiments with EPROMs*. In 2001, Dave won the Maggie Award for writing the best how-to article in a consumer magazine. To learn more about Dave's books and additional electronics projects, visit his Web site: www.pco2go.com.

INTRODUCTION
The Dawn of the Digital Lifestyle

Today's digital lifestyle actually began in the ashes of a B-29 incendiary bombing raid on Tokyo, Japan, during the nights of March 9 and 10, 1945. A young Imperial Japanese Navy officer toured the burned city and witnessed the destructive force that was capable of killing one hundred thousand souls in a couple of hours worth of bombing. These were the eyes of Akio Morita. Ironically, in a scant five years' time, this same city and this same man would give birth to one of the most incredible industrial stories of postwar Japan.

Like a phoenix rising from Tokyo's ashes, Morita and a fellow senior naval officer named Masaru Ibuka reunited to establish Tokyo Tsushin Kogyo K. K. (Totsuko) or Tokyo Telecommunications Engineering Corporation in a bomb-ravaged Tokyo department store. This reunion didn't happen immediately, however. Morita had to seek his father's permission to leave the family's lucrative sake brewery. Enlisting his father as a financial backer, Morita joined forces with Ibuka and helped inaugurate this new company on May 7, 1946.

In just one day after its inauguration, Totsuko had its first industrial order—50 vacuum tube voltmeters for the Ministry of Communications. The product was an immediate financial success for the fledgling company. This was a distinct turnaround from the previous year's product flop—the electric rice cooker.

Essentially, a wooden tub lined with some metal electrode plates, the cooker would either overcook or undercook the rice. Oddly enough, it was a combination of this success and failure that would lead to Totsuko's next big product.

Cloaked under a company pseudonym, Totsuko developed a bare-wire electrically heated cushion. It called the fake company Ginza Heating Company. Why the secrecy? Well, for one thing the electrically heated cushion wasn't protected with a thermostat. So overheating was a real problem. So much so that blankets, futons, and even trousers could catch on fire. Talk about a "hot seat." Also, Totsuko didn't want any public relations complaints from the electrically heated cushion to tarnish the technological merits of its other more sophisticated products such as the remarkable new record pickup known as "Clear Voice."

During the next couple of years, Totsuko expanded its factory operations. This remarkable growth necessitated two major relocation efforts. The first move was in mid 1946 to the Tokuya Building in Ginza. This was only a temporary

Sony = Precocious Boy

Just what does Sony mean? Well, according to Sony Corporation lore, the word "Sony" evolved from the combination of the Latin word "sonus" and the Japanese word "sonny." *Sonus* is a root akin to sound, whereas *sonny* is a term applied to an older brother or young boy. So, you might guess that by combining these two Sony cognates together, you would define a "noisy child," "loud kid," or a "brat." Not so fast, Noah Webster. Once again, returning to Sony Corporation history, the term "Sony" was coined to denote a small cadre of young men who were driven by energy and passion for unlimited creative inspiration. Although it might look like the *sonus* root has been conveniently ignored in this corporate marketing definition, remember that at the time of its inception, Sony had just produced Japan's first magnetic tape recorder and transistor radio. So sound was, and still is, an integral part of Sony.

host site for the sales operations of Totsuko. A greater, more urgent move loomed ahead.

In early 1947, Morita and Ibuka began scouring Tokyo for a site to host a combined sales and factory headquarters for Totsuko. The urgency for this move was dictated by the Kichijoji factory owner's eviction notice for Totsuko. Apparently, the telecommunications company was exceeding the electricity quota for the Kichijoji factory, and the owner feared having his service terminated. So, in February, Totsuko moved to Gotenyama in Shinagawa, Tokyo, on the site of a defunct Nippon Carburetor Company warehouse. This was a combined move with the sales force, corporate office, and production factory all housed under the same roof. Now Totsuko was poised for its first big consumer electronics product—the magnetic tape recorder.

> **CHEAT CODE:** From Akio Morita's autobiography, *Made in Japan*: "Our plan is to lead the public with new products rather than ask them what kind of products they want." And: "The public does not know what is possible, but we do."

The road to this landmark product wasn't a straight line, however. Initially, Totsuko engineers had developed a recording device that used wire for the recording medium. As archaic as this type of device sounds today, Morita was able to refine the design through the use of stainless steel wire, but the sound quality was still dubious. A chance encounter between Morita and Ibuka in the Civil Information and Education section of the Occupational Forces headquarters in the NHK building and an American tape recording device served as the catalyst for scrapping the wire recorder in favor of a magnetic tape recorder.

Totsuko released this new form of recording product in July 1950. It was more than just a reel-to-reel magnetic tape recorder; it was the first indigenous Japanese tape recorder. This event was quite an accomplishment for a nation that had been devastated by two atomic bomb detonations. In fewer than five years after the defeat of Imperial Japan, Morita and Ibuka were poised to become the owners of one of the world's largest consumer electronics manufacturers. Before that event would happen, though, Totsuko would have to become transistorized.

The patent rights to the transistor were owned by Bell Laboratories in the United States. Understandably, there was some objection from the Japanese Ministry of International Trade and Industry about purchasing the licensing rights to the transistor from Bell Laboratories. Effective marketing by Totsuko assured the Ministry that there was greater good to be achieved through this licensing venture. In fact, by August 1955, Totsuko had manufactured the first Japanese transistor radio. A major marketing success story derived from this contentious licensing agreement.

In a nod to the marketing savvy of Morita, Totsuko refined its transistor radio product, significantly reducing its size and coined the phrase "pocket-sized transistor radio." Unfortunately, Totsuko's pocket size didn't mesh with Morita's pocket-sized hype. So, in a stroke of design chicanery, Morita equipped all of his sales force with extra-large shirt pockets. Hence, the pocket-sized transistor radio was sold in supersized shirt pockets.

> **CHEAT CODE:** Sony Corporation's Design Division introduced two new colors, black and silver—black plastic parts and silver-colored metal. The first Sony product to sport this color design scheme was the TFM-110 "Eleven" FM radio.

Products such as the pocket-sized transistor radio helped transform Totsuko into a viable international consumer electronics manufacturer. With this global growth, however, the name Totsuko or Tokyo Tsushin Kogyo became difficult to assimilate into the world's rainbow of communities. Therefore, a major name change was ordered by Totsuko's upper management.

Attempting to forge an international corporate image, Totsuko was renamed Sony Corporation in January 1958. In the same month when the Soviet Union's Sputnik I burned up as it reentered the Earth's atmosphere, Sony Corporation rose as a global consumer electronics company. Just 12 years after its birth from Tokyo's fiery ruins, Sony Corporation began manufacturing a long succession of products that would become commonplace in every American's life. Names like Trinitron, Betamax, Walkman, Handycam, PlayStation®, VAIO, Memory Stick™, and AIBO filled U.S. homes with digital goodies promoting a new kind of living—a digital lifestyle.

> **CHEAT CODE:** Although thought of as "Mr. Sony," Morita didn't become Sony president until 1971, and, later, Chairman/CEO until 1976.

And Now PlayStation

Ask gamers about the year 1994, and they will get misty eyed about that fateful day in December when the Sony PlayStation was launched. Only a scant 10

FIGURE I-1
The Sony PSP handheld entertainment system. Take it out.

years would pass and the PlayStation Portable (PSP™) would be launched in Japan on the tenth anniversary of PlayStation's debut (December 12, 2004). Subsequent launchings in the United States (March 24, 2005) and Europe (September 1, 2005) would quickly catapult the PSP into a territory once dominated by Game Boy. A new consumer electronics niche was about to be carved out by Sony...again. (See Fig. I-1.)

If you thought that the video game industry was a "little kid" phenomenon enjoyed by a very small percentage of the adolescent population, think again. The video game industry is huge. How huge? Well, according to the NPD Group (an industry sales tracking organization), video game sales in the United States for the first quarter of 2005 were $2.2 billion. Now, let's put that figure into perspective.

Perfume or personal fragrance is a pretty big industry, right? An especially popular perfume product line is known as "celebrity fragrances." Once again, in a report published by the NPD Group, celebrity fragrance brand sales for 2004 were $94.9 million. So, playing a good video game is certainly more important than smelling like Elizabeth Taylor.

OK, how about we head north to Canada and study the apparel market? Clothes are big business and in a similar sales report from the NPD Group, the estimated dollar sales of the Canadian apparel market was $18.3 billion in 2004. Broken down into more specialized markets, these Canadian sales volumes were $10.2 billion for the women's apparel market and $5.7 billion for the men's market. Where did the other $4.4 billion sales go? Kids and specialty clothing lines rounded out the final $4.4 billion in sales. It would appear that a Canadian man might rather play a video game than wear clothes. Hey, hoser, take off, eh?

Now, compare the amount of Canadian women's apparel sales with the U.S. sales figures for women's sportswear cited by the NPD Group that amounted to $13 billion in 2004. In contrast, NPD Funworld, a division of the NPD Group that specializes in recording sales from the video game industry, reported that the U.S. combined sales of hardware and games (exclusive of computer games sales for both Mac and PC) for 2004 were $9.9 billion. Furthermore, of that combined sales figure, the Entertainment Software Association (ESA) reported that $7.3 billion of the total 2004 sales were attributed to computer and video game software sales alone.

> **CHEAT CODE:** In order to familiarize himself with American culture, Morita moved to New York City in 1963.

So why the comparison between video games and women's clothing and fragrances? Well, get ready, but according to an ESA fact sheet, in 2004, 43% of all game players were women. Want to hear another shocker? This same fact sheet stated that in 2004, 19% of Americans *over* the age of 50 played video games. And you wondered why game prices have climbed to nearly $50 a pop. Oh, and, in case you're also wondering how these sales figures relate to the number of actual units sold, the ESA fact sheet also claims that "more than 248 million computer and video games" were sold in 2004.

> **CHEAT CODE:** Akio Morita died on October 3, 1999, at age 78.

Taking all of this video game industry analysis one step further, a Nielsen Entertainment survey in 2005 cited two very important points. First, that 23% of U.S. households own a PC, game console, *and* handheld gaming device. Second, and more important, 57% of gamers own a Sony PlayStation 2 (PS2), whereas 39% own a Microsoft Xbox, and 27% own a Nintendo GameCube. As a footnote, 8% of gamers own all three types of gaming consoles. Now, that's my kind of household.

> **CHEAT CODE:** Not everything Sony touched turned to gold—in 1994, Sony announced a $3.2 billion loss from a failed investment in Columbia Pictures and Hollywood.

In spite of all of this sales survey hocus-pocus, it's not surprising that the venerable PS2 still reigns as the top video game console. Launched in March

2000, the PS2 is a flexible workhorse for Sony Computer Entertainment, Inc. (SCEI). In fact, SCEI released a statement on June 3, 2005, stating that the PS2 had reached 90 million units sold worldwide, as of June 2, 2005. Included in this lump figure were 36.48 million units sold in North America, alone.

This 90 million figure is a cumulative value that includes both the original PS2 and the newer, svelte PS2 SCPH-70000 series. Launched in November 2004, this new and improved PS2 version sold 16.17 million units by the end of March 2005. Oh, and in case you're keeping score at home, the original PlayStation sold more than 100 million units worldwide.

Play Unplugged

All of these sales figures lead back to one gaping hole in the Sony corporate gaming strategy—no portable game console. As early as the Electronic Entertainment Expo (E3) 2003, however, SCEI was making tentative marketing probes into the notion of a handheld gaming device. An ordinary portable game machine nee Game Boy wouldn't satisfy the SCEI designers, though. The Sony handheld had to be different. This revolutionary product would create a new market, not follow an existing market. Likewise, it had to have a signature design feature that would enable it to dominate this new market. But this device would also have to maintain a heritage with the rich and valuable PlayStation gaming market.

As Sony has demonstrated in the past, the best products are those that give consumers more than they expect. So, this new device would have to do more than just play games. It must also enable a multimedia experience. Music, photographs, and videos should also be compatible with this portable platform.

> **EASTER EGG:** The "other" Sony. Not every Sony product was a winner. There is an equally impressive list of product flops. Here is a short list of products that Sony might have wished had stayed on the engineer's drafting board:
>
> Electric rice cooker—there's nothing worse than bad food.
>
> Betamax—a proprietary format that wasn't licensed properly, Version 1.0.
>
> Discman—one man too many.
>
> Columbia Pictures Entertainment, Inc.—engineers make poor movie moguls.
>
> Mavica—these photos weren't worth one thousand floppy disks.
>
> VAIO—overpriced, overspec'd, bad name.
>
> Memory Stick—a proprietary format that wasn't licensed properly, Version 2.0.
>
> AIBO—an overpriced robotic dog that couldn't fetch a following.
>
> QRIO—tour de force robot with a Hollywood budget but lacking any wow-wee factor.

In the words of Kaz Hirai, president and CEO of Sony Computer Entertainment America (SCEA), extracted from an interview in *Official U.S. PlayStation Magazine*, April 2005 issue, Sony wanted to "...create a completely different market...using games as a catalyst." According to Hirai, the engineers at SCEI wanted to, "...make it a compelling entertainment option." Hirai continues to say that this device should enable you to, "...watch a movie..." and "...do a slide show... ." Also, Hirai states that this device must be a "music-playing device." As his wish list continued to grow, there was one thing that Hirai didn't want to have happen to this new product; have it end up, "...being a competitor to Nintendo or anybody else."

OK, fine, what should this new thing be called? Well, initially it was a "dynamic entertainment medium." (Note: This description still existed on the bottom of the initial PSP Value Packs sold in the United States.) Later, it became a "portable media player." During its launch in the U.S. market, SCEA coined the term "portable entertainment system." This moniker lasted a scant two weeks, when SCEA adopted the more apropos "handheld entertainment system." And, thus the PSP handheld entertainment system was formally

What's a PSP?

The original U.S. PSP model featured an impressive array of technical specifications:

- → 4.3-inch, wide-screen, high-resolution LCD
- → High-capacity (1.8-gigabyte) optical disc UMD drive
- → Memory Stick Duo port
- → USB 2.0 port
- → IR port
- → Wi-Fi 802.11b wireless LAN
- → Stereo speakers
- → External headphone connector
- → Analog joystick
- → Directional button keypad

FIGURE I-2 **A brilliant use of branding: The back of the PSP tells the entire world exactly what you're holding, playing, and watching in your hands.**

crowned. (See Fig. I-2.) There is a curious footnote to this PSP branding effort. Initial PSP advertisements used the phrase "Portable Entertainment Revolution." These ads were glossy, printed card-stock sheets that included perforated, life-size replicas of the PSP that could be punched out by prospective buyers. Supposedly, owner wannabes could then marvel at the amazing dimensions of the PSP. Ironically, some stores had stacks of these cardboard PSPs languishing on shelves as late as August 2005.

Right out of the chute, the PSP came out with its guns blazing. Within two days of sales in the United States, the PSP sold 500,000 PSP Value Packs. The figures for sales of the PSP alone exceeded $150 million in the first week on the U.S. market. This wasn't a flash in the pan, either.

The NPD Group's video game sales data for May 2005 reported that the PSP sold 250,000 units, whereas the Nintendo DS and Game Boy Advanced (GBA) combined to sell about 279,000 units. Several unsubstantiated reports circled the Internet stating that the Nintendo sales figures were further broken down as: Nintendo DS sold approximately 57,000 units, and GBA sales totaled 223,000 in May. This breakout could *not* be confirmed with the NPD Group, Nintendo, or Sony.

Enter the Digital Lifestyle

So, how do you ensconce a handheld entertainment system into a new market niche? Well, in the case of iPod, Apple Computer imbued its white plastic music player with a hip cachet that also insulated the user from every technical aspect of digital music management. It didn't hurt this marketing tack either that the iPod sported a revolutionary technological feature—the scroll wheel. Now the ball was in Sony's court.

Rising to the challenge of surmounting a similar marketing obstacle, SCEA trumped the iPod's scroll wheel with a stunning, full-color, 4.3-inch, widescreen (16:9), liquid crystal display (LCD) screen. (See Fig. I-3.) Furthermore, the innovative silhouette iPod advertising campaign was replaced by runways lined with fashion designers while high-profile celebrities and well-known fashion models strutted their stuff appointed with PSPs. Real people using real PSPs; real cool. The Sony PSP had become a fusion of technology with fashion.

On March 14, 2005, SCEA converted the outdoor plaza of the Pacific Design Center in West Hollywood into the PSP Style Park. This exclusive fashion accessories show was known as *Prêt à PSP*. Many one-of-a-kind PSP accessories were created by a healthy list of great designers. Marc Jacobs, Diane von Furstenberg, Henry Duarte, Coach, LRG, Jennifer Lopez, Baby Phat by Kimora Lee Simmons featuring Simmons Jewelry Company, Heatherette, Autore Pearls, C. Ronson by

FIGURE I-3 You can't tell by looking at these three PSPs, but each one is markedly different from the other. Each system features a different version of the PSP firmware. From left to right: 1.50, 1.51, and 2.0. See Chapter 7 for a complete explanation of these software differences and how they can dramatically affect your game play.

Charlotte Ronson, (wb) waraire boswell, Jenni Kayne, Kidada for Disney, and IceLink Watch each contributed some remarkable baubles, all of which tantalized an adoring crowd of celebrities.

Although the silhouette figures in iPod advertising might offer a certain feeling of creative ambiance, it's tough to ignore the raw animal magnetism that can be derived from half a dozen super models strutting down a runway with a Sony PSP hanging around their necks. Models Maggie Rizer, Kirsty Hume, Guinevere, Rhea Durham, Alek Wek, Nicky Hilton, and Bijou Phillips echoed the sleek and sexy lines of the PSP while providing a glimpse ahead into the upcoming fall fashion accessory lineup. This was fashionable technology.

Rushing to fill their messenger bags after the show with prerelease PSPs (the U.S. PSP launch date wasn't until 10 days later), David Arquette, Courtney Cox Arquette, Drea de Matteo, Wilmer Valderrama, Jason Bateman, Vince Vaughn, Owen Wilson, Nicky Hilton, Jessica Alba, Regina King, Chris Kattan, Alyssa Milano, Paula Abdul, Adrian Grenier, Kevin Dillon, Kevin Connelly, and Jeremy Pivon used the onsite Sony store in the Access Boutique to enjoy the PSP experience.

The digital lifestyle was beginning ahead of schedule.

Oh, and if you are interested in appointing your PSP with some of these fashionable accessories, then you should consider the pure gold case encrusted with seven carats of yellow and black diamonds by Baby Phat by Kimora Lee Simmons featuring Simmons Jewelry, or maybe the tan suede muff with rabbit fur trim and gold studs created by Jennifer Lopez. Of course, basic black is always an excellent choice, and the black leather multipocket clutch with leather wrist strap by Marc Jacobs is the perfect accent to every gamer fragging the brains out of an opponent.

Enter Your Digital Lifestyle

That was then, this is now. Right? Well, actually you can purchase your own custom-made finely crafted leather case for your PSP, right now. Cases from Norêve and Vaja are excellent choices for the discriminating PSP owner. Or, if gaming-on-the-go is your passion, then the indestructible cases from Otterbox will keep your PSP dry and protected while you're surfing the waves and surfing the Web. Likewise, there is a vast assortment of PSP cases for the proletariat—the working gamer's case. But there's more to the PSP accessory market than just a good case.

Skins, batteries, screen protectors, headphones, grips, amplifiers, chargers, removable media, USB cables, and sunscreens are just a sampling of the optional extras that will heighten your PSP experience. Not all is golden in the PSP accessory market, however. For example, how do you tell which is the best skin for personalizing your PSP? Well, you're holding the answer in your hands, right now.

This book will help you separate the gems from the junk. Every worthwhile PSP accessory is thoroughly and painstakingly reviewed and critiqued. Armed with the knowledge contained in this book, you will be able to outfit your handheld entertainment system with the finest quality accessories that are currently available. Furthermore, by using some of my secret industry contacts, I was able to get the inside scoop on some of the future accessories that haven't even hit the market, yet. Therefore, you can get the drop on your gaming companions and have your PSP outshine all others.

There's more to the PSP than accessories, however. There's movies, photos, and music. Plus, the PSP is rooted in the one of the richest, most exciting game lineups in the history of video games—the Sony PlayStation game library. It's not just a matter of buying the best PS2 game for your PSP, though. Some games are better suited for the handheld environment. But which ones? Once again, you're in luck. This book looks at every PSP game and evaluates its portable playability. Huh? Well, some games are better suited to being played on the big screen rather than on a smaller one. For example, although *Gran Turismo 4: The Real Driving Simulator* is a fantastic PS2 game, the PSP incarnation can become slightly claustrophobic. Therefore, this title didn't make it into this book's Top 10 Games listing in Appendix B.

More than 75 PSP games were evaluated for this book. You can study this evaluation in Chapter 7. There's more to gaming than the big-name games that are sold on Universal Media Disc (UMD™). There is another market that has sometimes been erroneously lumped into the same category as crackers and computer software thieves. This market is commonly referred to as "homebrew" games. First, I must get on my soapbox: **I personally will not report nor condone the copyright infringement activities sponsored by some cracker elements of the homebrew market.** So, don't expect to

> ### A Value Deal
>
> The initial PSP Value Pack (with a suggested retail price of $249.99) contained the following items:
> - PSP (PlayStation Portable)
> - AC adaptor/charger
> - Battery pack
> - Memory Stick Duo™ (32 MB capacity)
> - Earbuds with remote control (strangely molded in white plastic)
> - Soft case
> - Wrist strap (another odd although leather accessory)
> - Cleaning cloth
> - UMD game/movie/music sampler
> - *Spider-Man™ 2* from Sony Pictures Home Entertainment (only included with the *first one million* PSP Value Packs shipped)

read about this type of activity in this book. This book does, however, thoroughly acknowledge efforts at installing and playing legitimate and legal homebrew games on the PSP. Who could argue about the inventive merits of playing a demo version of *DOOM* via a homebrew interface on the PSP? It can be done and you can read about how you can do it in Chapter 7.

But the PSP is also more than games—it's a handheld entertainment system. As such, unlike any other product on the market, the powerful gaming resources of the PSP have been brilliantly leveraged for access to a large and growing library of feature-length motion pictures. Take that Game Boy; take that iPod.

These movies are only available on UMDs. Unlike Betamax, the UMD format is being embraced by many different motion picture studios. Therefore, the selection of UMD movies is continuing to expand and evolve. Whereas the initial UMD movie offerings were high on testosterone, the emerging market is changing to offer a wider range of cross-sectional theatrical productions. So *Hellboy* now sits next to *Hitch*. And that's a good thing.

Games Available for PSP

U.S. Launch

Ape Escape®: On the Loose, SCEA

ATV Offroad Fury®: Blazin' Trails, SCEA

Darkstalkers Chronicle™: The Chaos Tower, Capcom

Dynasty Warriors®, KOEI

FIFA 2005, Electronic Arts

Gretzky™ NHL®, SCEA

Lumines™, Ubisoft

Metal Gear Acid™, Konami

MLB™, SCEA

MVP Baseball, Electronic Arts

NBA, SCEA

NBA Street Showdown, Electronic Arts

Need for Speed™ Rivals, Electronic Arts

NFL Street 2 Unleashed, Electronic Arts

Rengoku™: Tower of Purgatory, Konami

Ridge Racer™, Namco

Smart Bomb, Eidos Interactive

Spider-Man 2™, Activision

Tiger Woods PGA TOUR®, Electronic Arts

Tony Hawk's Underground 2 Remix, Activision

Twisted Metal: Head On™, SCEA

Untold Legends: Brotherhood of the Blade, Sony Online Entertainment

Wipeout® Pure, SCEA

World Tour Soccer, SCEA

All told, during the preparation of this book more than 110 movies were viewed on my PSP. As you can imagine, some videos made the transition to the smaller screen elegantly, although others were poor half-pint versions of their bigger DVD™ brethren. How do I know? Well, I compared both the DVD *and* the UMD version of every PSP movie. I then compiled a no-holds-barred listing of the Top 10 PSP movies. You can read my assessment in Appendix C.

You don't have to limit yourself to watching UMD movies, though. You can, "roll your own," if the creative spark ignites your filmmaking talents. Unlike the self-contained UMD movies, however, loading your movies onto your PSP takes a little bit of effort. In Chapter 1, several great how-to tutorials guide you through *all* of the steps necessary for getting your music, photos, and movies on your PSP. Even better, I'll show you how to make your own music videos. Although I can't guarantee that your efforts will be broadcast on MTV, I can assure you that you will certainly attract a sizable crowd gathering around your PSP at the airport. Oh, and Chapter 1 is completely platform nonspecific, so that both Mac and PC owners will be able to load their personal content on a Memory Stick for premiering on their own PSP.

UMD MOVIES RELEASED ON APRIL 19, 2005, FROM SONY PICTURES HOME ENTERTAINMENT
- *Hellboy*
- *House of Flying Daggers*
- *Once upon a Time in Mexico*
- *Resident Evil 2: Apocalypse*
- *XXX*

Finally, if modding, hacking, or tricking out your PSP sound appealing, there are several chapters that are perfect for satisfying your do-it-yourself (DIY) self. For example, learn how to recharge your PSP from your computer while you're connected to its USB port. Or, add electroluminescent wiring to the exterior of your PSP for some dazzling visual effects. And I can't forget to tell you about Chapter 4 where I'll show you how to use older Memory Sticks with your PSP.

That's not all, though. There's six other chapters and couple of extra appendices, too. Way more information that you will probably need right now. As your tastes change and your PSP experience increases, however, this book will grow right along with you. Your PSP is not only a handheld entertainment system, it's the foundation of a revolutionary way of looking at gaming and entertainment. It's your digital lifestyle, so get ready to live it.

CHAPTER 1

Yeah, But How Do I...?

Welcome to the digital revolution. May I please see your admission ticket? Good, you brought your Sony PSP with you. Everything seems to be in order; your admission is granted. Please enjoy the rest of this book. You are about to embark on your new digital lifestyle. Be sure to fasten your seat belt, this ride could be a bit bumpy.

As the owner of the world's premiere handheld entertainment system, you have no doubt mastered some of the finer elements of basic PSP operations. For example, you've

- → Inserted the battery pack
- → Charged your battery
- → Turned on your PSP
- → Initialized your system language, time zone, and date and time settings
- → Selected and entered a nickname for your PSP
- → Turned off your PSP

THE FINER POINTS OF THE PSP ANATOMY

PSPS-1 Power up your PSP with this power switch on the right side.

PSPS-2 One of the most valuable ports on your PSP is the USB port. Refer to Chapter 10 to learn how to fully exploit this port.

PSPS-3 Your PSP has wireless network connectivity via this Wi-Fi switch on the left side.

PSPS-4 Recharge your PSP battery with this charge connector.

PSPS-5 Pelican makes a car charger for playing your PSP on the road. Just make sure that you keep your eyes on the road. No, really.

PSPS-6 Most modern automobiles are equipped with special power outlets that can accept the Pelican charger.

PSPS-7 Games and movies for your PSP are sold on this type of removable UMD media.

PSPS-8 Slide this UMD release switch on the top side of your PSP for loading UMD media.

PSPS-9 Oddly enough, the UMD media faces backwards when inserted into your PSP. Weird.

PSPS-10 Slide the UMD media in and snap this UMD disk drive cover closed.

PSPS-11 Your PSP can also use removable Memory Stick PRO Duo media cards. Open this cover to load a media card into your PSP.

PSPS-12 Just like the UMD media, the media card faces backwards when inserted into your PSP. Weird, again.

PSPS-13 Push this media card into your PSP until it clicks into place.

21

Now what, right? Well, if you have to ask that question, then we need to talk. Can you spell "games," "movies," "music," and "pix"? Those are exactly the types of things that you should be enjoying on your PSP. If you're ready to partake in these finer digital lifestyle delicacies, then please skip to the next section in this chapter. Otherwise, if you are currently PSP-challenged, then you can discover how to unleash your PSP's potential and learn everything that you need to know about this incredible handheld entertainment system in the following Top 10 PSP Questions and Answers section.

Top 10 PSP Questions and Answers

1A. QUESTION: My battery seems to lose its charge even when I'm not using my PSP. Why?

ANSWER: Are you kidding me? You're not using your PSP *all* of the time? Jeez, watch a movie, play a game, listen to some music, use your PSP. But, I'm getting ahead of myself. Your handheld entertainment system uses power (just a trickle), even when it is off. Therefore, charge your PSP regularly. Also, monitor your battery charge level with the icon on the PSP system screen. In addition to regularly charging your PSP, you should also strive to never let the system operate for extended periods of time at its lowest charge levels (Fig. 1-1).

FIGURE 1-1 Use the "Battery Information" screen for monitoring your battery's performance.

1B. Q: Well, how long will my battery last?

A: Hmm, that's a little tougher question. First of all, every battery pack is different. Likewise, various games and movies draw varying levels of power during PSP operation. Want more? OK, here are some real world measurements about battery strength, endurance, and charging, not a bunch of hyped manufacturer flimflam:

> **CHEAT CODE:** At one time the menu system of the PSP was called the XrossMediaBar™ or XMB™.

At a 75% battery charge level (as indicated with the "Battery Information" setting) on a four-month-old battery, the movie *Hellboy—Director's Cut* (approximately 132 minutes in length) could be watched in its entirety. At the conclusion of the movie, the battery charge level was 28%. The recharging of this battery then took 1 hour 55 minutes, and the resultant battery charge level was 100%.

2. Q: Why is my screen so dark? I can barely see anything on it, and I can't use it outdoors.

A: Did you know that Sony has officially cautioned users about *not* using the PSP outdoors while driving an automobile? Doh. Seriously, the intensity or brightness level of the PSP can be controlled with the Display Button. There are three levels of brightness that can be toggled by repeatedly pressing the Display Button. Alternatively, you can manually turn off the display by holding the Display Button down for approximately one second. As soon as you press any other button on the PSP, the screen will reappear magically, like a glowing golden nugget emerging from the muddy water at the bottom of a gold-rush prospector's pan. Or, something like that.

3. Q: I don't see what the big deal is, the PSP will never replace my iPod. Why doesn't my music sound very good on my PSP?

A: Could it possibly have something to do with your taste in music? The PSP is an extremely competent music player. Although Linear PCM and ATRAC3plus formats can be played on the PSP, most users will opt for listening to the ubiquitous MP3 format music. No matter what your file format preference, you can adjust the music's tonal output with the Sound Button.

There are six settings for the Sound Button:

SETTING	TONE	HOW TO ACCESS
HEAVY	Intense bass	Press button
POPS	Better vocals	Press button
JAZZ	Balanced range	Press button
UNIQUE	Wide spectrum	Press button
OFF	Normal tone	Press button
MUTE	Off	Press and hold button for approximately 1 second

But here's the kicker. *None* of these tone settings can be adjusted *without* the headphones attached to your PSP. Furthermore, altering the tone settings *during* game playing will have *no* effect. The PSP is set to a default Off or normal tone while playing games.

> **CHEAT CODE:** You can obtain more detailed battery charge level information with the "Battery Information" setting inside the "System Settings" option from the "Settings" menu. Got that?

4. Q: I want to watch DVD movies on my PSP. How do I insert a DVD into my system?

A: Oh, you've got to be kidding? Did you try a crowbar? Really, the PSP can play DVD-quality movies, but *only* from *UMD format discs*. Let's back up a bit here.

Digital video disc or DVD is an optical disc storage technology. Think of DVD as a CD-ROM disc on steroids—DVD can hold cinema-quality video, CD-quality audio, still photographs, and computer software. The DVD has become one of the few bridge products that is able to span home entertainment, gaming, and business markets—all on a single disc format. All is not bliss in DVD, however. Slowly, many different storage formats have populated the DVD medium. And, in doing so, the simplicity of the cross-market DVD has given rise to an acronym mess.

- DVD-ROM—the base storage format
- DVD-R—recordable format
- DVD-Video—movie base format; aka DVD
- DVD-Audio—music base format
- DVD-R/RW—recordable, writable format

- DVD-RAM—removable storage format
- DVD+R/RW—write once, writable format
- DVD-VR—video recording format
- DVD+RW—writable video format
- DVD-AR—audio recording format
- DVD-SR—stream recording format

> **CHEAT CODE:** You can limit the maximum volume on the PSP by activating the Automatic Volume Limiter System with the "Sound Settings" setting inside the "System Settings" option from the "Settings" menu.

And, I would be remiss if I didn't mention the PS2 DVD-ROM format for video games.

Did you notice something missing from that alphabet soup mess? There was no mention of UMD.

Universal Media Disc or UMD is a proprietary medium that is capable of delivering digital entertainment content (e.g., games, movies, and music) to your PSP. This high-capacity optical medium is described by Sony as "the next-generation compact storage media." Measuring a scant 60 millimeters in diameter and weighing about 10 grams, each UMD can store up to 1.8 gigabytes (GB) of digital data. And contrary to what you might have been lead to believe, this disc's capacity is ample for holding a full-length feature motion picture.

> **CHEAT CODE:** Turn off the WLAN feature prior to charging your battery. The battery cannot be recharged while this feature is in use.

Technically speaking, the UMD features 128-bit AES data encryption along with MPEG4 AVC, ATRAC3plus Caption PNG video format codecs and ATRAC3plus, PCM, (MPEG4 AVC) audio codecs. The result is a glorious 16:9 widescreen movie display. In summary, UMD video for the PSP features DVD picture quality in a wide-screen presentation. As such, you can *only* play UMD videos in your PSP.

In June 2005, SCEI, the game unit of Sony Corp, announced that its UMD had been approved as a standard format by Ecma International (formerly European Computer Manufacturers Association). This announcement opened up the PSP game and movie markets. Even the giant arm of Sony's music group, Sony Music Entertainment, began embracing the UMD format. This widespread acceptance isn't without some unfortunate fallout, however.

One little nuance of the UMD that is being exploited by game manufacturers is the requirement that newer games *must* have the latest and greatest update for the PSP system's firmware. In other words, you can't play the UMD

> **EASTER EGG:** Confused about how to create folders and file structures on a Memory Stick PRO Duo? Just format your Memory Stick with the PSP's built-in Memory Stick Format function. You can access this function with the "Memory Stick Format" setting inside the "System Settings" option from the "Settings" menu. This format function will create separate folders for music, photographs, and saved game information. It will not create the folders hierarchy needed for video, however. Also, please remember that formatting a Memory Stick PRO Duo will erase all data currently on the media card. So use your newfound power responsibly.

game that you just purchased, unless you upgrade your handheld entertainment system's firmware. Granted this upgrade is usually included on the UMD (or freely available via the Internet), but the reasoning behind the required update is distressing—upgrade or you *can't* play the game.

This mandatory upgrade mentality is particularly upsetting for those PSP owners who wish to run homebrew software. You see, these firmware updates are attempts by SCEA to "plug" holes in the system software that hackers are able to exploit for coding and running homebrew games. OK, enough background, here's how to use a UMD with your PSP.

Now, to play a UMD (game or movie) in your PSP, you must hold your handheld entertainment system upright while releasing the Open Latch that is located on the top of the PSP. The Disc Cover will pop open. Hold the UMD with the printed label facing the PSP's back panel and insert the disc into the UMD tray until it is completely seated inside the PSP. Remarkably, the printed label will be *upside down* when inserted properly. Go figure, eh? Also, be sure that you don't touch the disc's read opening that is located on the underside of the UMD. When the UMD is properly seated, just snap the Disc Cover closed.

Just reverse these steps for ejecting the UMD. Hold the PSP upright, slide the Open Latch, remove the UMD, and close the Disc Cover. One final caution is in order: *Never* eject a UMD during game or movie playback. Also, don't play in traffic and always eat your vegetables.

5. Q: My Memory Stick won't fit in my PSP. Why not?

A: Ah, the joys of proprietary formats. Yes, you may have a Memory Stick, but it's not the right Memory Stick. Right?

→ Memory Stick—Standard-sized media with copyright protection and high-speed data transfer.
→ Memory Stick PRO—Standard-sized, advanced media with insertion guides.
→ Memory Stick Duo—Compact media for mobile devices including copyright protection. An adapter is available for using this media with standard-sized Memory Stick compatible devices.
→ Memory Stick PRO Duo—Compact media for high-speed mobile devices. An adapter is available for using this media with Memory Stick PRO compatible devices. This format is compatible with the PSP.

Note that,

1. Memory Stick PRO and Memory Stick PRO Duo media cannot be used to record ATRAC3plus (including ATRAC3) format music files.
2. Recording with MagicGate copyright protection (Digital Rights Management) functions can only be performed in MagicGate compatible products.
3. Management software consumes a portion of Memory Stick media memory storage space. For example, approximately 5 megabytes (MB) of media space is used on a 128-MB Memory Stick leaving approximately 123 MB of space for your game data, movies, photographs, and music.

> **CHEAT CODE:** Never remove a Memory Stick PRO Duo card while the Memory Stick Duo access indicator LED is flashing.

The flavor of Memory Stick that is used in the PSP is the Memory Stick PRO Duo. There are two primary sources for these removable flash-memory media cards: Sony and SanDisk®. As far as quality and performance, both of these brands are virtually identical. In terms of price, however, the SanDisk Memory Stick PRO Duo cards are your best buy (Fig. 1-2).

Now available in a fancy translucent color-coded package, all SanDisk Memory Stick PRO Duo media have large capacity (e.g., 128 MB to 2 GB), low power, fast transfer, and very reasonably priced. Both MP3 music and MPEG4 video

FIGURE 1-2
SanDisk Memory Stick PRO Duo media cards are your best buy for use in your PSP.

FIGURE 1-3
SanDisk ImageMate 12-in-1 reader/writer.

FIGURE 1-4
Look, Ma, no drivers. Finally, plug-and-play that really works.

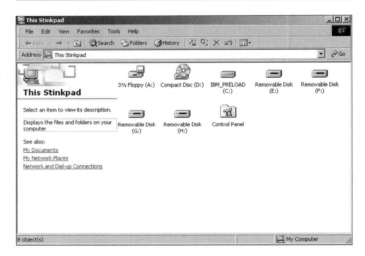

benefit from these features. Rather than writing your content to the Memory Stick PRO Duo card via the PSP's USB port, a better solution is to use either the SanDisk ImageMate 12-in-1 reader/writer (Fig. 1-3) or the SanDisk MobileMate for Memory Stick reader/writer. Both of these products connect to your computer's USB port (e.g., Mac or PC), require no supplemental operating system software drivers, and receive their power via the USB port (Fig. 1-4). Neat.

Although the MobileMate is specialized for reading and writing all four Memory Stick media, the ImageMate 12-in-1 reader/writer can use the following memory card formats:

- CompactFlash Type I
- CompactFlash Type II
- SD
- miniSD
- MMC
- RS-MMC
- Memory Stick
- Memory Stick PRO
- Memory Stick Duo
- Memory Stick PRO Duo
- SmartMedia
- xD

> **CHEAT CODE:** You can create playlists for your PSP. Simply make a .M3U file listing your preferred play sequence for the music tracks and copy this file into the specified music folder. When you select this M3U file, your playlist will play as a musical track group on your PSP.

There is a lot more to this Memory Stick issue. Please read Chapter 4 for more information on how to use other Memory Stick media with your PSP.

Once you have the correct Memory Stick media, it's a simple three-step process for using these cards:

1. Turn the PSP over and lift up the Memory Stick PRO Duo Slot Cover.
2. Grasp the Memory Stick PRO Duo card and push it into the slot until it is fully seated. The Memory Stick PRO Duo label should be facing you. There is a direction arrow on the label side of all SanDisk cards that serves as a visual cue to help you with this process. Likewise, there is a notched corner that will give you a tactile cue for which end goes in first.
3. Close the Slot Cover.

When you want to remove a Memory Stick PRO Duo card, just open the Slot Cover and press the card. This action will disengage the spring that holds the card in place and eject the Memory Stick PRO Duo. Remember to close the Slot Cover when you are done removing/replacing the media card.

6. Q: My pix stink on my PSP. How can I make my photos shine?

A: First of all, don't include so many photographs of yourself. Next, you have to make sure that your selected imagery is up to snuff with the PSP's JPEG rendering engine. And, the best tool for all digital image editing is the venerable Adobe Photoshop.

Because navigating Photoshop can be a dark and mysterious maze unto itself, just follow along with this PSP-specific procedure for manipulating a photograph into perfect proportions for PSP viewing.

1. Select a suitable digital photograph. I have selected the adorable Amelia playing with her beloved Robosapien.
2. Open the image in Photoshop.
3. Select the Crop tool and crop the photograph's dimensions to 480 × 272 pixels. These dimensions represent the size of the PSP screen. If your photograph is unable to accept these dimensions, then the resulting JPEG will be surrounded by a black border on the PSP. The size of this border will depend on the difference between your photograph's dimensions and 480 × 272 pixels.
4. Monitor the dimensions of your cropping action with the dimensions window.
5. Complete the cropping action by double-clicking inside the cropped selection of the photograph.
6. Save a copy of this image as a JPEG. Select the maximum image quality option and make it a standard JPEG.

Once you have a properly dimensioned JPEG image, you are ready to load it on a Memory Stick PRO Duo media card. In order for this image to be displayed by your PSP, all photographs must be saved in the proper location. This location involves a funky file and folder hierarchy structure that you can either

FIGURE 1-5
Automatically create your PSP folder hierarchy by letting your PSP format your Memory Stick PRO Duo media card for you.

"roll" yourself or let your PSP create (i.e., with the "Memory Stick Format" setting) the correct hierarchy (Fig. 1-5). In either case, your photographs must be copied into this folder:

Memory Stick PRO Duo -> PSP -> PHOTO -> YourFolderName -> Your Photos

Your photographs can now be viewed, zoomed, rotated, and panned while viewing them on the PSP. A terrific slideshow function can also be implemented for displaying all of the photographs inside a select folder. You can adjust the speed of the slideshow's transition (e.g., normal, fast, or slow), by selecting the corresponding option from "Photo Settings." Best of all, you can control all of these photograph options with the PSP system buttons. For example, click the L button to display the previous photograph or click the R button to display the next photo. Just a couple of clicks and you can scroll through your entire photo album.

POWER USER MEGA-HACK. If you want to gain the maximum flexibility and versatility out of your PSP image viewing capability *and* use the outstanding hacks cited in Chapters 4 and 10, then you should copy *all* JPEGs into a folder named DCIM (Fig. 1-6). *Don't* use the PSP -> PHOTO folder hierarchy.

FIGURE 1-6
Be a PSP power user; don't put your pix inside the PHOTO folder. Instead use the DCIM folder labeled "Digital Camera Images."

Inside this DCIM folder you must also include another folder named after a digital camera or other digital media device's folder. For example, if you use Olympus camera products, you might name this folder 100OLYMP (i.e., that's 100 and OLYMP). Or, you could use the original folder name sponsored by Sony for Memory Stick cameras: 100MSDCF. Regardless of this folder's exact name, just drop your JPEGs into this digital media device's folder and you can view them on your PSP as "Digital Camera Images." Furthermore, you can copy these images to iPods, external USB hard disk drives, USB Flash Memory drives, cameras, printers, and so on. In fact, this Power User Mega-Hack will make your PSP into a universal media player that even Sony never envisioned.

Note that,

1. Folder and filenames are case insensitive.
2. You can't have any folders within the *YourFolderName* folder.
3. A folder named Digital Camera Images might exist inside the PHOTO folder. This is a PSP system folder reserved for photographs loaded directly from Memory Stick Duo-equipped cameras. This feature is a neat trick if you currently own a digital camera that uses Memory Stick Duo media cards for recording photographs. Just take the Memory Stick Duo out of the camera and pop it into the PSP. Scroll to the Photo icon on the main menu and select the "Memory Stick" option.

7. Q: Every time I load my MP3 music onto my PSP, I see twice as many files as I have album tracks, and half of them are labeled as "Incompatible Data." What's the problem?

A: Whether it's a problem or not, I don't know, but your symptoms sound like you are using a Mac to move your music files onto your PSP. The files that you are seeing are actually invisible files that are useful to the Mac OS X (i.e., all versions of new "jungle animal" Mac OS), but worthless to all other systems; including your PSP. These invisible Mac filenames typically begin with a period (.) followed by the filename. For example, one of the most infamous invisible Mac files is .DS_Store. This file is everywhere on any removable volume (e.g., server, hard drive, USB device, Memory Stick, PSP, etc.) that touches your Mac OS X system. Although this is a bit of sloppy engineering, these files are totally harmless to all other systems including your PSP.

Now, if you want to get rid of all of these invisible files, you have two choices. First, use a commercial application for copying your photographs, music, and videos onto your PSP. Applications like PSPWare for Mac OS X from Nullriver Software insulate you from all of the worries about *some* invisible files. Even if you elect to use these commercial applications, every time you jack into your PSP with a Mac, you will insert these sloppy .DS_STORE (and maybe even .TRASH file) files onto your handheld entertainment system.

If you're like me and you want to take matters into your own hands, you can manually clean all invisible files from your Memory Stick PRO Duo media. You just need some patience and the free Mac OS X application, Terminal.

You can find Terminal inside the Utilities folder that is in your Applications folder. Double-click the Terminal icon and you are greeted with an austere command line interface window. Nothing fancy here—no icons, no pretty colors; just type in UNIX commands and behold the beauty of raw unbridled power. Once you have Terminal up and running, just follow these simple steps for ridding your PSP of invisible files:

1. Depending on how your PSP is connected to your Mac (via USB or a SanDisk Memory Stick reader/writer), you might need a slightly different file navigation scheme, but basically you want to use the change directory (cd) command to get to your PSP media. For example, when I connect to my iBook via a PSP USB connection, I can use this command for navi-

HOW TO CUSTOMIZE YOUR PHOTOGRAPHS FOR THE PSP

PS-1 **Start with a great photograph. In this case, it's Amelia and Robosapien.**

PS-2 **Open your photo in Adobe Photoshop (or, use your fav photo editor) and select the Crop tool.**

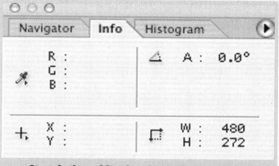

PS-3 Stretch the rubber band Crop box around your photo until the box measures 480 × 272 pixels. You can monitor your pixel pickin' with the Info tab.

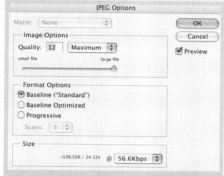

PS-4 Save your JPEG as a standard baseline image with maximum quality.

PS-5 Copy all images into the PHOTO folder inside the main PSP folder.

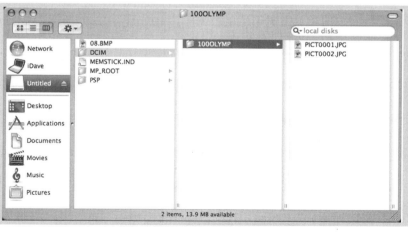

PS-6 Be a power user hacker. Don't put your photos into the PHOTO folder. Instead use the DCIM folder for holding all of your PSP images.

gating to the Memory Stick PRO Duo media: **cd ../../Volumes/Untitled**. "Untitled" is the name of the PSP Memory Stick PRO Duo media that my Mac OS X assigned. Your media's name *could* be different. Just look on the desktop to see what name your Mac assigns when you plug the PSP or media reader/writer into your computer.

2. Now I keep carefully stepping through the PSP's folder hierarchy structure: **cd PSP**.
3. And **cd PHOTO**.
4. Finally, I'm in the correct folder, **cd Family\Pix**. Notice that I had to use an "escape" nomenclature for telling Terminal that my folder name has a "space" in it. In this case, the folder's name is: "Family Pix." To show UNIX that there is a space between "Family" and "Pix," I need to define the space with the backslash keystroke. Therefore, the space is represented as "\".
   NOTE: Yes, you can stack all of these directory changes into one command line operation: **cd ../../Volumes/Untitled/PSP/PHOTO/Family\Pix**. You power user, you.
5. Also, if at any time you want to see what the folder contents look like during any of these operations, just use the list files/folders command and include the "all" switch. In other words, **ls -a** will show you the entire contents of the current folder—both normal files and invisible files.
6. Now the biggie, the remove command (rm). This type of file deletion is not like today's wimpy operating systems where you will be prompted ad nauseam about whether or not you really want to delete a file—in Terminal "rm" will delete a file—immediately. I want to delete the .DS_Store invisible file, so I type this command: **rm .DS_Store**. And, it's gone.
7. Those "Incompatible Data" music files are more prevalent than the solitary .DS_Store file. So, a little power user trick must be employed for speeding up the file removal process. In this case, a wild card is used with the remove command. A wild card is a command substitute for a specific character or group of characters. The "*" wild card is used to represent a group of characters. By using this wild card, I can remove all of the invisible "Incompatible Data" files with one simple command: **rm ._*.*** .
8. Close Terminal, the job is done. If you have other invisible files hiding on your PSP, just navigate to the contaminated folder and remove the offending files, too.

Rock on.

8. Q: I've tried hundreds of ways to copy video onto my PSP, but nothing works. How do I convert DVDs and movies to work on the PSP?

A: This is the tale of two tricks. First, you have to convert your DVD or movie to a format that works with the PSP. Then, you have to copy the converted video to the proper location on your PSP.

In answer to the first part of this video trickery, there are several applications that are available for turning this operation into a simple drag-and-drop solution.

- PSP video converters
- 3GP Converter 031
- ffmegX
- PSP Video 9
- PSPWare
- Sony Image Converter 2
- X-OOM Movies on PSP™

> **CHEAT CODE:** You can select a fourth brightness setting (the maximum brightness setting allowed) by plugging the AC adapter into the PSP.

One of the newest kids on the PSP-video-converter block is *X-OOM Movies on PSP*. Although all of these other video converters are good products, there are two features that are available with *X-OOM Movies on PSP* that make it a significant product worthy of serious consideration.

First of all, *X-OOM Movies on PSP* is able to convert and encode commercial DVDs to work on your PSP. I will not address the copyright infringement implications of this capability, but I do feel that the inclusion of this feature is a terrific option that *all* PSP owners should embrace.

Want an even better reason to buy *X-OOM Movies on PSP*? It's called X-OOM Fit-to-Stick technology. What this means is that you tell *X-OOM Movies on PSP* the size or capacity of your Memory Stick PRO Duo media card and the software's encoder will compress and resize the final movie so that it fits on your stick. Isn't that awesome? No more hoping that the final movie will fit.

The truly sad thing about *X-OOM Movies on PSP* is that it will only work on a PC. Too bad for Mac owners; *X-OOM Movies on PSP* is a great PSP product. I must mention this caveat regarding this program: It did have some severe dif-

HOW TO DELETE MAC OS INVISIBLE FILES FROM YOUR PSP

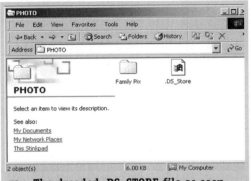

TS-1 The dreaded .DS_STORE file as seen with Microsoft Windows 2000. In the Mac OS, this file is invisible.

TS-2 This naughty Mac "._" file is an invisible file that is flagged on the PSP as "Incompatible Data."

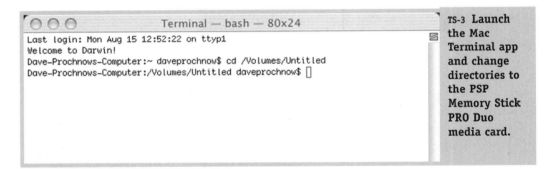

TS-3 Launch the Mac Terminal app and change directories to the PSP Memory Stick PRO Duo media card.

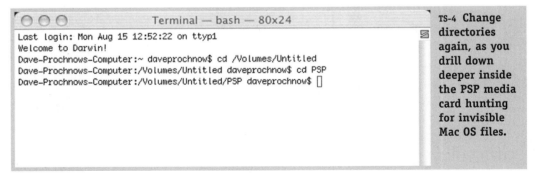

TS-4 Change directories again, as you drill down deeper inside the PSP media card hunting for invisible Mac OS files.

```
Last login: Mon Aug 15 12:52:22 on ttyp1
Welcome to Darwin!
Dave-Prochnows-Computer:~ daveprochnow$ cd /Volumes/Untitled
Dave-Prochnows-Computer:/Volumes/Untitled daveprochnow$ cd PSP
Dave-Prochnows-Computer:/Volumes/Untitled/PSP daveprochnow$ cd PHOTO
Dave-Prochnows-Computer:/Volumes/Untitled/PSP/PHOTO daveprochnow$ cd Family\ Pix
Dave-Prochnows-Computer:/Volumes/Untitled/PSP/PHOTO/Family Pix daveprochnow$ 
```

TS-5 **You're going to have to use some special tricks to navigate past folder names that have spaces in them.**

```
Last login: Mon Aug 15 12:57:50 on ttyp1
Welcome to Darwin!
Dave-Prochnows-Computer:~ daveprochnow$ cd /Volumes/Untitled
Dave-Prochnows-Computer:/Volumes/Untitled daveprochnow$ cd PSP
Dave-Prochnows-Computer:/Volumes/Untitled/PSP daveprochnow$ cd PHOTO
Dave-Prochnows-Computer:/Volumes/Untitled/PSP/PHOTO daveprochnow$ cd Family\ Pix
Dave-Prochnows-Computer:/Volumes/Untitled/PSP/PHOTO/Family Pix daveprochnow$ ls -a
.               ..              ._Amelia.JPG    Amelia.JPG
Dave-Prochnows-Computer:/Volumes/Untitled/PSP/PHOTO/Family Pix daveprochnow$ 
```

TS-6 **Now let's see those pesky invisible files.**

```
Last login: Mon Aug 15 12:57:50 on ttyp1
Welcome to Darwin!
Dave-Prochnows-Computer:~ daveprochnow$ cd /Volumes/Untitled
Dave-Prochnows-Computer:/Volumes/Untitled daveprochnow$ cd PSP
Dave-Prochnows-Computer:/Volumes/Untitled/PSP daveprochnow$ cd PHOTO
Dave-Prochnows-Computer:/Volumes/Untitled/PSP/PHOTO daveprochnow$ cd Family\ Pix
Dave-Prochnows-Computer:/Volumes/Untitled/PSP/PHOTO/Family Pix daveprochnow$ ls -a
.               ..              ._Amelia.JPG    Amelia.JPG
Dave-Prochnows-Computer:/Volumes/Untitled/PSP/PHOTO/Family Pix daveprochnow$ rm ._*.*
Dave-Prochnows-Computer:/Volumes/Untitled/PSP/PHOTO/Family Pix daveprochnow$ ls -a
.               ..              Amelia.JPG
Dave-Prochnows-Computer:/Volumes/Untitled/PSP/PHOTO/Family Pix daveprochnow$ 
```

TS-7 **Zap the invisible files and return to your desktop. Not a bad day's work for an invisible file zapper.**

ficulties reading/converting some movie file formats. For example, an IBM ThinkPad operating Windows 2000 and running *X-OOM Movies on PSP* was unable to decode several advanced systems format (ASF), audio video interleave (AVI), and MOV (a file extension used by QuickTime-wrapped files) files that had been compressed with various codecs. Although, in the case of the MOV files, PSPWare was able to decode all of them on a Mac when *not* using QuickTime. Go figure.

Basically, no matter what the name of your poison is, the use of these video conversion applications is all the same:

1. Install the software.
2. Select the conversion format; PSP video, MPEG4, and PSP MP4 are common examples.
3. Drag the source video file onto either the application's icon or its conversion panel.
4. Go grab a cup of coffee; big movies take a long time to convert.

Now that you have your converted video in hand, err, on your hard drive, it's time to master the second tricky riddle in the PSP video puzzle. Getting your video from point A to point B when you can't even C it.

Optionally, some applications will copy the final PSP MPEG4 files directly to a Memory Stick PRO Duo media card that has been properly mounted on the computer prior to performing the video conversion. Otherwise, you will have to manually copy the converted files onto your PSP. The destination for your videos is inside a folder named 100MNV01 that is located in the folder MP_ROOT. This MP_ROOT folder is in the root directory of the PSP Memory Stick PRO Duo media card. It does *not* go inside the PSP folder. Therefore, the complete path for your videos is: MP_ROOT/100MNV01.

There are two files that are created from each movie that you convert into a PSP video MPEG4 format: the video file and a still image file, named MP4 and THM, respectively. The MP4 file is the MPEG4 video conversion of your original movie. Each file conversion application automatically names the final MP4 file with this naming convention: M4Vxxxxx.MP4, where the xxxxx portion is a unique number. Similarly, the still image is actually a JPEG file that is extracted from the opening frame of the original source movie. This JPEG file is given a THM extension (thumbnail) and a name that matches the MPEG4 file's name: M4Vxxxxx.THM. This thumbnail is the image that you see inside the "Memory Stick" option of the "Video" menu on the PSP main menu.

If you are creative and want to give your PSP videos some flair, you can use Adobe Photoshop for making your own thumbnails. Just take a JPEG image that will make a suitable movie billboard and modify its image size to a dimension of 160 × 120 pixels. Save your new thumbnail and be sure to change the .JPG extension to .THM. Also, your thumbnail's filename must exactly match the MPEG4 video's name. Now copy the thumbnail into the MP_ROOT/100MNV01 folder and enjoy.

9. Q: I've read all this talk about WLAN and PSP Wi-Fi. What is Wi-Fi and how do I install it on my PSP?

A: Lucky you: Wi-Fi is already installed on your PSP...and everyone else's PSP, for that matter. Wireless fidelity or Wi-Fi is a popular buzzword established by the Wireless Ethernet Compatibility Alliance (WECA) for representing the IEEE 802.11b communications protocol. Just like ethernet is a fast platform-nonspecific protocol for file sharing through wired connections, Wi-Fi is a fast platform-nonspecific protocol for file sharing through wireless connections.

When WECA certifies a product as being Wi-Fi compliant, that product is able to communicate with any other Wi-Fi product regardless of operating system and manufacturer. So your Wi-Fi PSP is able to communicate with an Apple Airport Extreme Base Station. No wires, no problems. Neat, eh?

Although it is wonderful that the PSP is Wi-Fi compliant, setting up your network connection can be a daunting task. Not an impossible task, just a lot of little steps that must be performed very accurately. Like so many things in life, though, the first step is the easiest. Turn on the wireless local area network (WLAN) switch on the left side of the PSP (Fig. 1-7). Done. Now, the remaining steps are performed within the "Network Settings" option on the "Settings" menu.

XS-1 *X-OOM Movies on PSP* is a commercial package that will convert video content into PSP movies.

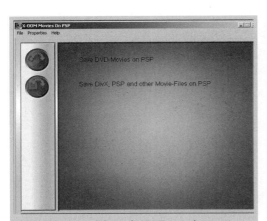

XS-2 You can either rip DVD movies onto your PSP or convert video files for running on your PSP.

XS-3 There isn't much in the way of optional settings for *X-OOM Movies on PSP*.

XS-4 **Ripped DVD movies can be exported directly to the Memory Stick media inside your connected PSP.**

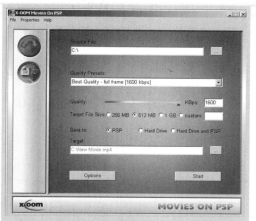
XS-5 **Similarly, video files can be converted and saved to an attached Memory Stick media. You can also determine the final PSP MPEG4 file size.**

XS-6 **A status panel monitors the movie conversion process.**

HOW TO CONVERT YOUR VIDEO INTO PSP MOVIES

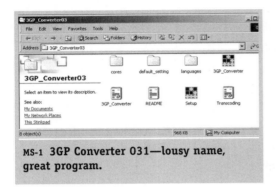

MS-1 **3GP Converter 031—lousy name, great program.**

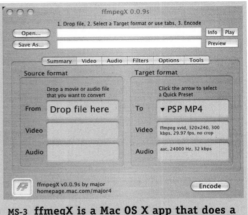

MS-3 **ffmegX is a Mac OS X app that does a mediocre job of movie file conversion.**

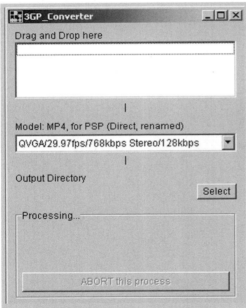

MS-2 **3GP uses a simple drag-and-drop interface. In fact, you can just drop your movies files on the desktop icon, and 3GP will do the rest.**

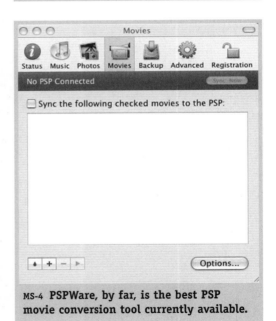

MS-4 **PSPWare, by far, is the best PSP movie conversion tool currently available.**

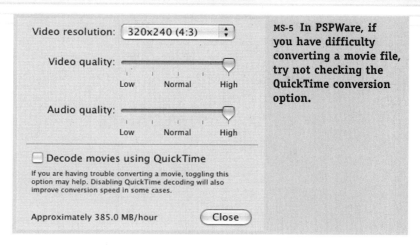

MS-5 In PSPWare, if you have difficulty converting a movie file, try not checking the QuickTime conversion option.

MS-6 MPEG4 movie files and their respective thumbnail images are ready for viewing on this Memory Stick PRO Duo media card.

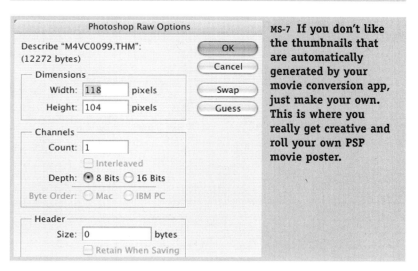

MS-7 If you don't like the thumbnails that are automatically generated by your movie conversion app, just make your own. This is where you really get creative and roll your own PSP movie poster.

FIGURE 1-7 **The WLAN switch on the left side of the PSP.**

FIGURE 1-8 **You can set up your ad hoc WLAN configuration in the "Network Settings" option.**

FIGURE 1-9 **You can either let your PSP determine its own ad hoc channel, or you and your fellow fraggin' friends can go completely CB and wire in your own channel.**

There are two types of WLANs: ad hoc (Figs. 1-8 and 1-9) and infrastructure. Which one should you use? If you want to connect to another PSP or PSPs that are within the same room, your UMD game will automatically enable the ad hoc WLAN for you. That's it—start gaming. Software that uses ad hoc WLAN might have some additional configuration information.

Choose the infrastructure WLAN when you want to connect to a network or the Internet. If you elect to go online with your PSP, make sure that you have the following network items:

→ An ISP
→ A wireless router or WLAN access point

Granted, the ISP part of the equation is simple enough, but what type of wireless router should you consider? Without hesitation, I would recommend the Apple Airport Extreme Base Station (or Airport Express). Simply stated, this wireless router is the easiest, most powerful wireless network system that you can buy. I have used it with terrific success with the PSP. Before you buy one, though, you should know that you will need at least one Mac OS wireless computer for configuring Airport Extreme Base Station. Yes, the system will function seamlessly with PCs, but you still need an Airport-equipped Mac for setting up the system.

No matter what brand of wireless router that you use, you will need some information from it for connecting the PSP to your Wi-Fi compliant network:

→ Service Set Identifier (SSID)
→ Wired Equivalent Privacy (WEP) Encryption Type
→ WEP Key

What the heck is all of this stuff? First of all, SSID is just the 32-character (or less) name of your WLAN. For example, if you set up your wireless router with the name "Wi-Fi 1Der," then your SSID would be "Wi-Fi 1Der." A scan feature is built into the PSP for helping you locate and identify all available WLANs within reach of your PSP. This feature is located inside the "Network Settings" option of the "Settings" menu.

The WEP is a security protocol developed for wireless networks for encrypting data during transmission. There are two elements that are needed by the PSP for entry into your WLAN SSID. First of all, you must know the type of encryption used for your network's WEP. The most common forms of encryption are 40-bit and 128-bit. Some wireless access points such as Internet cafés even use no WEP encryption.

Once you know the type of WEP encryption used on your WLAN, then you must obtain the WEP key for accessing this network. You can consider this key to be like a password for gaining access to the network. Unlike most passwords that you're used to, the WEP key is a hexadecimal (hex) password. And, it can be really long. Here's an example: The WLAN named **Wi-Fi 1Der** uses 128-bit WEP with a key that is 13 characters long. On the wireless router, this key is entered as: **planEtEthEr23**. This is *not* the key that is entered into your PSP. Rather, the hexadecimal equivalent password WEP key for this WLAN would be: **0BCAB922E485E880B279D00B99**. (Note: This is a simulated hex WEP key.) This *is* the WEP key that you would type into your PSP. Have fun keying that hex sequence into your PSP.

After you have configured your PSP and saved your settings, you can test your connection. Select the "Test Connection" option inside the "Network Settings" and wait for the verdict. If you see this response, **Error Code: "A connection error has occurred. 80110482"**, then you probably made a mistake typing in your lengthy WEP key. Don't fret, it's an easy mistake to make. Just try again. If you still get an error message, try changing the settings of your WLAN system; try 40-bit encryption or no encryption. Actually, during my lengthy experimentation with the PSP and establishing wireless network connections, I found that 128-bit WEP encryption with a 13-character key worked best for my Airport Extreme Base Station.

You can learn more about wireless communication in Chapter 9.

10. Q: I read somewhere that my PSP has a remote control. Where is my remote control?

A: Probably around your neck. Although it doesn't look like much, that little round disk that is attached to your PSP earbud headphones is actually a multi-function remote control (Fig. 1-10). In fact, the remote control will function differently, depending on the entertainment activity that you are currently enjoying.

OK, the most obvious time to use the remote control is during listening to your music (Fig. 1-11). In this capacity, you can increase/decrease volume, play, pause, skip, and fast forward/reverse your music selection. Simple enough. Switch to watching a video and the operation of the remote control changes, too.

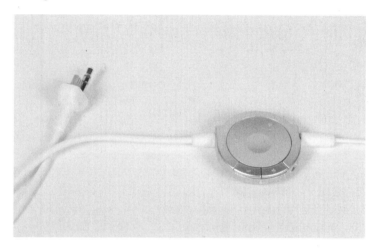

FIGURE 1-10
PSP remote control.

FIGURE 1-11
Plug the remote control into your PSP line out jack.

Using the remote control while watching a video you can play, pause, skip to chapters, and fast forward/reverse portions of the video (Fig. 1-12). Be forewarned, not all of these functions work with videos that have been stored on Memory Stick PRO Duo media cards. For example, you can't skip to different chapters on videos that are viewed from media cards.

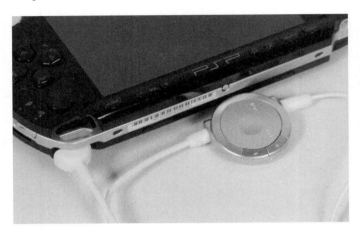

FIGURE 1-12 **The remote control acquires different functions based on whether you are listening to music, watching a movie, or viewing pix.**

When you view the photographs stored on your PSP, the operation of the remote control changes, again. The remote control will now play a slideshow, pause a slideshow, display the previous image, and display the next image.

All of this variety from the same little silver disk (Fig. 1-13). A chameleon might wish to be so versatile.

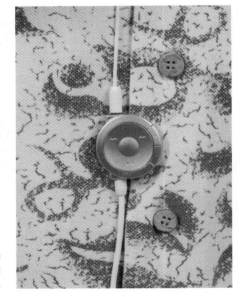

FIGURE 1-13 **Just clip the remote control on your shirt and controlling your PSP is just a lapel away.**

50   CHAPTER 1

CHAPTER 2

The Hack Heard 'round the World

I had to giggle as I read the April 5, 2005, CNN.com article titled "Hackers Add Web, Chat to PSP." I was reading the article on my PSP. As one of the first owners of a PSP in late March, I was also able to exploit the *Wipeout® Pure* (also spelled WipEout PurE) SCEA Web site browser in just 5 hours of tinkering. Soon, the entire Internet was buzzing with information for turning this fun, futuristic racing game into a potent Web browser. Wait a minute, it isn't that good of a Web browser, and most of the claims about this hack's prowess were sheer hyperbole.

This hack is a finicky and pathetically slow interface for browsing the Web. It will, however, work in a pinch for grabbing a quick (err, slow) headline from a Wi-Fi hot spot. Say, for example, while you're waiting for a flight at an airport. Originally designed as an interface for enabling players to download new ship designs, known as "race craft," and additional track layouts from the *FX300 Racing League*, this hack modifies the domain name system (DNS) lookup of the SECA Web site that supplies this downloadable content and enables you to point your PSP to another DNS server. If you're still interested; great, let's get started.

This is a deceptively easy hack to perform, and it will *not* harm your PSP. Before you begin, make sure that you have enabled your PSP to connect to a

CHEAT CODE: Want to start your own *Wipeout Pure* Web site portal? The first step is that you must have a static IP address. Most Web sites that are housed on a shared system don't have a static IP. In other words, if you "rent" Web space from a typical Internet service provider (ISP), you probably don't have a static IP at your disposal. There are some ISP services, however, that will allow you to purchase a static IP for your Web site. If this is the case, then you can purchase the IP address and set up a Web page for handling *Wipeout Pure* page requests.

Once you have your static IP, you will need to add two things to your Web server. First, you will need a folder named: wipeout. Inside this folder you will need to add an HTML file named: index.html. This is the Web page that will function as your *Wipeout Pure* Web portal.

FIGURE 2-1
Don't forget the Wi-Fi. If your PSP Wi-Fi switch isn't on, then you can't get online.

Wi-Fi network (Fig. 2-1). Read Chapter 1 for more information on establishing this type of connection. Don't take this precaution lightly, if you don't have a solid, working Wi-Fi connection, this hack will be very difficult to complete.

Make Your Mark on the World (Wide Web)

1. Select the "Network Settings" option on the "Settings" main menu.
2. Choose the "Infrastructure Mode" option.
3. Select an established (and working) Wi-Fi connection. Leave all of your entered connection information intact. Scroll through each of the various screens until you reach the "Address Settings" screen.
4. Choose the "Custom" option from the "Address Settings" screen.
5. On the "IP Address Setting" screen, select the "Automatic" option.
6. On the "DNS Setting" screen, select the "Manual" option.
7. You will be entering your selected static Internet protocol (IP) address as the "Primary DNS" entry. Leave the "Secondary DNS" entry as: 0.0.0.0.
8. On the "Proxy Server" screen, select the "Do Not Use" option.
9. Review your settings and save your settings by pressing the "X" button.

> **CHEAT CODE:** The official *Wipeout Pure* Web site is: www.wipeoutpure.com. Use this Web site for downloading supplemental game data.

> **EASTER EGG:** Rather than purchasing a static IP, if you would rather operate the portal for *your* own benefit, then go local. To establish a portal on your home network, you will need the following basic setup:
> 1. A wireless router with the capability of assigning an IP address to your local computer. This address becomes the DNS entry for your PSP *Wipeout Pure* hack.
> 2. A Web-sharing setup on your local computer.
> 3. Create a wipeout folder inside your shared computer's directory. Write an index.html file for the wipeout folder. This file will be displayed on your PSP when you select the "Download" entry from the *Wipeout Pure* "Main Menu."

CHEAT CODE: If you would like to try and track down the elusive *Wipeout Pure* download Web site with your browser, you will encounter some interesting information. Begin your sleuthing with this URL: ingame.scea.com/wipeout/index.html. Nose around and you will discover that SCEA is using an Apache/1.3.27 server. Watch your step; too much nosing around could get you this kind of access: **FORBIDDEN: YOU DON'T HAVE PERMISSION TO ACCESS /WIPEOUT/INDEX.HTML ON THIS SERVER.**

10. Make sure that you've inserted the *Wipeout Pure* UMD into your PSP. Now, scroll across to the "Game" menu and select "UMD." Start *Wipeout Pure* and select the "Download" entry on the "Main Menu." The PSP will request a wireless network connection from you. Select the Wi-Fi connection that contains the new DNS settings from Step 7. Your selected Web portal should now be on your PSP screen (Fig. 2-2).

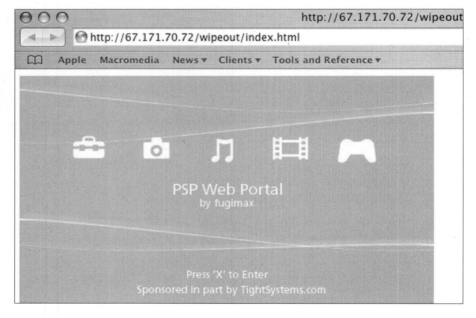

FIGURE 2-2 An example of one of the many PSP Web portals that sprang up when the *Wipeout Pure* hack was released.

Enjoy your new 16:9 view of the world.

All of this changed when Sony released Firmware 2.0 for the PSP in the United States on August 24, 2005 (Fig. 2-3). Or, as Gilda Radner would say, "Never mind."

FIGURE 2-3 **An Internet Web browser was built into Firmware 2.0. This feature was an extremely attractive reason for upgrading your PSP and risking the loss of support for running homebrew applications.**

CHAPTER 3

SAW: The PSP Dissected

Are you ready to void the warranty of your PSP? To know the inner workings of your handheld entertainment system, you're going to have to take the screws out and open it up. Right? No? You don't feel comfortable sawing your personal digital lifestyle into pieces?

OK, I'll do it for you. Unlike the UMD movie with a similar title (*SAW*, 2004, Twisted Pictures), you will not be forced to make a choice between keeping your precious PSP intact or hacking it into pieces. Rather, in this chapter, I will guide you through the whole gory process with a painless illustrated navigation of the PSP's anatomy.

> **CHEAT CODE:** The grandfather of game console hacking is the thirty-something Benjamin J. Heckendorn. This kid's got a future.

Please don't be squeamish. And, please pay attention. The lessons that you learn in this chapter will come in very handy during some actual hacking and modding later in this book. Hey, you can't expect to trick out your PSP without getting under the hood. Can you?

I've been "under the hood" of five different PSP systems. In each instance, I've been surprised by one consistent element that is found on *all* PSPs—Phillips-head screws. Yes, Sony could have used some tamperproof screw for the PSP; rather, it elected to put the ubiquitous DIY screw in its revolutionary

CHEAT CODE: You can easily draw a steady +3 volts from the PSP to drive your own hacks, mods, and projects.

game machine. What does that mean? Well, if you read it like a hacker, the use of Phillips head screws means that the PSP was designed to be hacked. So let's get to work.

First, a Word from My Lawyer

Hello, I'm Dave's lawyer. Before you attempt to dissect your, or anyone else's, PSP, please read, understand, and accept the following warnings, precautions, and disclaimers regarding the disassembly of a PSP. Thank you.

PRECAUTIONS

Disassembling the PSP will void your warranty. There is no authorization for the disassembly or modification of the PSP. There could be a risk of electrical shock or fire by disassembling the PSP.

WARNINGS

The LCD screen contains dangerous, high-voltage parts. Always remove the battery prior to disassembling the PSP (Fig. 3-1).

FIGURE 3-1 Remove the PSP battery before you attempt to disassemble the system.

That's an Unbelievable Display of...

Without a doubt, the Thin Film Transistor (TFT; aka active-matrix display) LCD screen on the PSP is the most stunning feature on this handheld entertainment system. Remarkably, inside this Sony product, it is a Sharp LCD that is responsible for all of this hubbub. This LCD is Sharp part number LQ043T3DX01. Don't expect to find a specifications document for this LCD on the Sharp Web site, either. Contrary to popular misinformation, this Sharp LCD is *not* an off-the-shelf part. Rather, it is a special production part that was specifically designed for meeting Sony's performance requirements. Therefore, you're out of luck for finding any exact interface specifications. What you can find, however, is a very close approximation LCD made by Sharp.

The Sharp part number LQ038Q5DR01 is a 3.8-inch color TFT-LCD display module that is designed primarily for the automotive industry. Considered a small TFT LCD, the LQ038Q5DR01 (www.sharpmeg.com/part.php?PartID=628) is very close to many of the specifications for the PSP Sharp LQ043T3DX01. Luckily, there is a specifications document for the LQ038Q5DR01 that can help shed just a little light on the PSP LCD.

According to the specifications document for the LQ038Q5DR01, this 320- X 240-pixel module consists of the LCD panel, driver integrated circuits (ICs), control board, frame, shielding, and backlight. The panel driver is a 40-pin ribbon cable that plugs into a KX15-40K-type connector (Fig. 3-2). A secondary 2-pin ribbon cable drives the backlight fluorescent tube and mates into a SM02(8.0)B-BHS connector. So, what does all of this gobbledegook mean? Basically, that the PSP LCD consists of two connectors that plug into the main circuit board. That's it. So yes, we're close to finding a way to display PSP imagery on a larger LCD, but we aren't close enough. Yet!

Oh, and in case you're wondering, you can purchase the LQ038Q5DR01 from a variety of electronics parts sources. Digi-Key Corporation (www.digikey.com), for example, sells the LQ038Q5DR01 for approximately $403 each. All of a sudden, your $250 PSP looks very attractive, doesn't it?

> **EASTER EGG:** Would you like to get a slightly used **PSP** to use as your victim, err, test subject for hacking, modding, and experimenting? First, locate an independent game store in your area. Shy away from the big-name consumer electronics chain stores. Instead, find a local shop that is owned and operated by a local entrepreneur. Once you've located a store, ask the owner if there are any opened or demo **PSPs** that he or she would be willing to sell for a reduced price. Many owners would love to part with a smeary, scratched, smudged **PSP** and replace it with a brand new shiny one. If you're lucky enough to find a dealer who will accept your offer, you could end up saving yourself enough money to buy a new **UMD** movie.

DISCLAIMERS

SCEA, The McGraw-Hill Companies, and Dave Prochnow will neither assume nor be held liable for any damage caused to anyone or anything that is associated with the disassembly of the PSP. The warranty for the PSP will be considered null and void if the warranty seal on the PSP has been altered, defaced, or removed.

Scalpel...

Our patient is ready; let's see what makes this sucker tick.

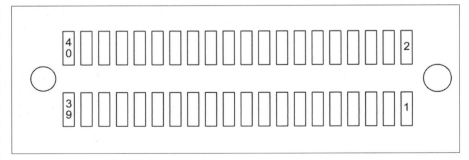

FIGURE 3-2 The pin out for the LQ038Q5DR01 KX15-40K-type connector. Notice how the pin assignments alternate between top and bottom.

HOW TO DISASSEMBLE YOUR PSP

SS-1 Turn the PSP over onto its face. Protect the screen with the Sony-provided polishing cloth.

SS-2 Make sure that you remove the battery before you attempt to disassemble the PSP.

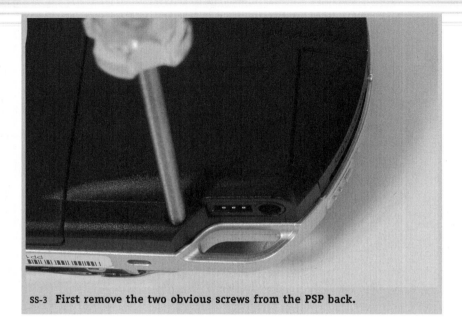

SS-3 First remove the two obvious screws from the PSP back.

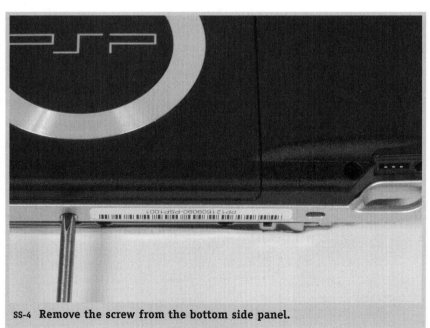

SS-4 Remove the screw from the bottom side panel.

SS-5 **Now you're about to void your warranty. Make sure that you accept this prospect and then remove both labels from the battery compartment. Yes, the screws are only under the warranty label, but I cleaned out the entire compartment for future mods and hacks. You might want to record your serial number in your instruction manual before you remove the big upper label.**

SS-6 **Remove the battery compartment screws. Also, remove any media cards or discs that are inside the PSP at this time.**

SS-7 Carefully flip the PSP over onto its back and lift the front case cover off.

SS-8 Wrap this front case inside the polishing cloth for safe keeping.

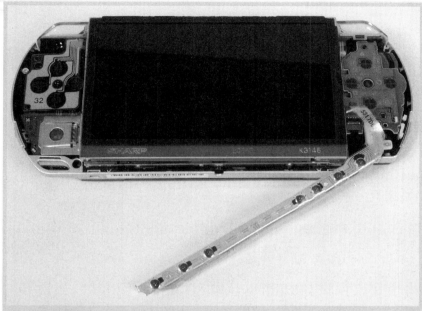

SS-9 **Disconnect the lower button bar.**

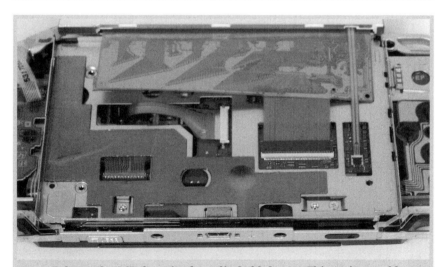

SS-10 **Release the LCD from its four clip hold-downs. This action could take a little *careful* and *gentle* prying with a flat screwdriver blade.**

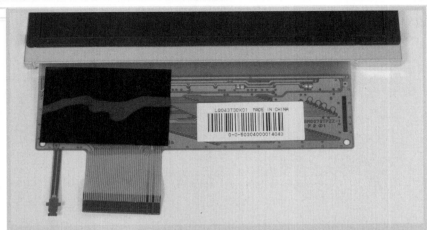

SS-11 Unclip both the backlight and the LCD driver ribbon cables from the PSP and remove the LCD. These cable connectors have small, *delicate* snap covers that must be lifted up prior to removing the cable. Try using a small wooden toothpick for this operation.

SS-12 The LCD driver and backlight connectors.

SS-13 ...place with a single screw.

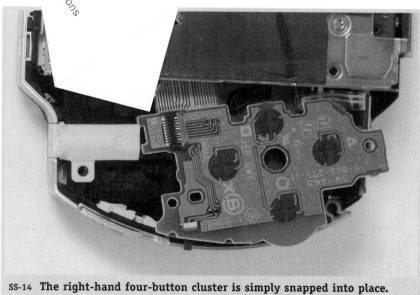

SS-14 The right-hand four-button cluster is simply snapped into place.

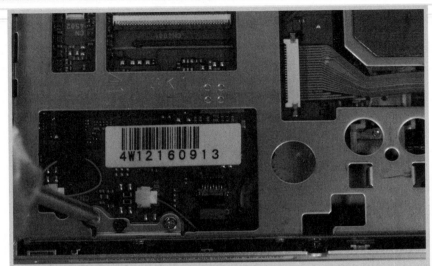
SS-15 Begin removing all of the screws that hold the metal shielding/LCD bracket onto the main circuit board.

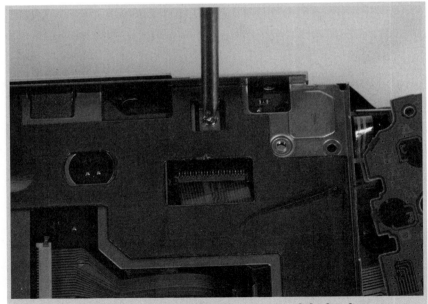
SS-16 There are six screws located in three corners of the bracket.

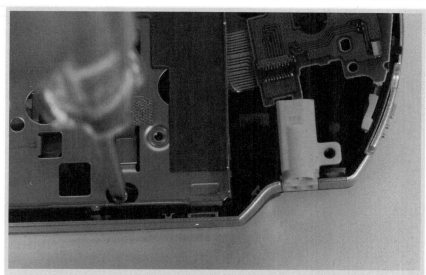

SS-17 Remove the metal shielding/LCD bracket. Use extreme caution as you remove this bracket—it also holds the latch mechanism for the UMD drive door.

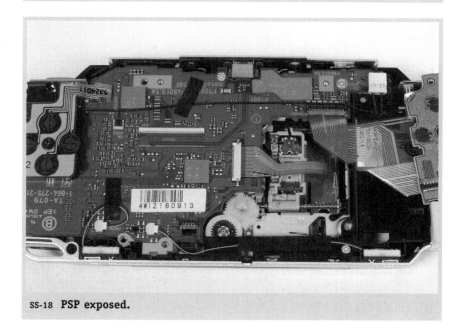

SS-18 **PSP exposed.**

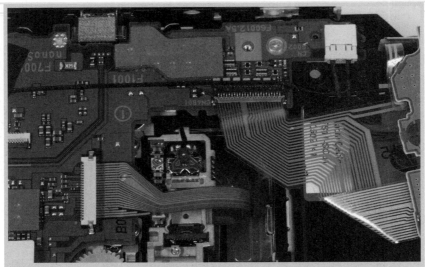

SS-19 The other side of the UMD drive. This connection is another area that is ripe for hacking.

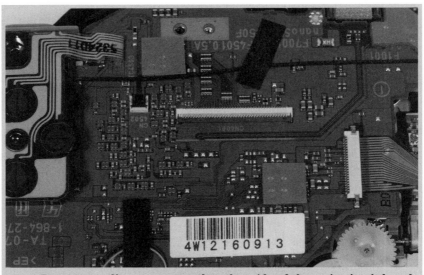

SS-20 For extra credit, you can see the other side of the main circuit board —the side that I can't show you. All you have to do is follow the previously described steps and then remove the main circuit board.

SS-21 And Humpty Dumpty *was* back together again. In fact, as further testimony that the PSP was designed to be hacked, reassembly is a snap, and it should work when you power it back up. I have never had a DOA PSP. Yes, I've smoked a couple, but I've never failed to resuscitate a reassembled PSP.

CHAPTER 4

Shtick Talk

Many people erroneously believe that because the Memory Stick is a proprietary storage format that only Sony products use it. Au contraire. There are many products and many manufacturers that have embraced the Memory Stick format.

One great fallout from this widespread use is that you can find a lot of Memory Stick products for hacking into PSP accessories. And, that leads us to an easy PSP hack: a hack that enables you to use your older Memory Stick media cards with your PSP.

For me, this hack started back at my local cell phone company. I was just window shopping for some of the latest and greatest cell phone technology, not really interested in anything in particular, just looking for innovative products that might be fun to hack.

In the corner, where last year's technology is pushed aside for selling the hot new phones, there was a little blue-gray phone from Sony Ericsson (Fig. 4-1) that caught my eye. After only a quick study, I could tell that this was a remarkable phone.

FIGURE 4-1 Sony Ericsson strives to instill innovation and feature-rich designs in the mobile handset industry. (Photograph courtesy of Sony Ericsson Mobile Communications AB)

Packed inside its little 4.6- × 2.3- × 1.1-inch frame was a "candy bar" form factor unlike any that I had ever seen. There was an MP3 player, digital camera, MPEG4 video player, USB support, personal organizer, Java™ game system, and Bluetooth™ technology all squeezed into a cell phone. Oh, but it gets better. This cell phone supported removable media—Memory Stick Duo media cards, to be precise. Good golly, Miss Molly. Actual, factual multimedia messages could be fabricated on the fly from Memory Stick files, digital camera images, streaming video, and enhanced messaging. This was the Sony Ericsson P800 (Fig. 4-2).

> **CHEAT CODE**: MagicGate™ is a digital rights management (DRM) technology that attempts to safeguard copyrighted material on Memory Stick media cards. Basically, MagicGate works by first authenticating a Memory Stick as a legitimate MagicGate device, then encrypts the contents. To streamline this transfer process, the host computer encrypts the data (e.g., music, images, videos, etc.) with a contents key, while the Memory Stick encrypts a session key. Each Memory Stick has a unique storage key that is used by the host for generating this contents key. Later during playback, the session key is then used by legitimate hosts to decrypt the media contents.

FIGURE 4-2 **The Sony Ericsson P-800. (Photograph courtesy of Sony Ericsson Mobile Communications AB)**

Nestled next to the P800 was a funny little device that looked like someone has just tossed it there. After a little discussion with the always-friendly sales clerk, I found out that this quirky accessory would enable a P800 phone to accept the larger, older, and cheaper Memory Stick media cards in the Memory Stick Duo slot on the phone. Whoa, Catherine Zeta-Jones, I had to have this thing. Still, not knowing much about either the P800 or this accessory, I purchased it and brought it home.

> **CHEAT CODE**: You can learn about the Memory Stick specifications in Shigeo Araki, "The Memory Stick," *IEEE Micro*, Vol. 20, No. 4, IEEE Computer Society Press, Los Alamitos, Calif., 2000.

The packaging for this accessory was distinctly bland. There was no documentation; just a cryptic two-step pictorial that showed the Memory Stick adapter attached to a P800 cell phone. Luckily, there

That's My Memory and I'm Sticking to It

This is just a sampling of the many different products that use the Memory Stick. Not mentioned are the large number of SanDisk Memory Stick products. Please refer to Chapter 1 for an in-depth look at these SanDisk products.

Digimax U-CA 4—Samsung Techwin
Cyber-shot DSC-L1—Sony Corporation
Cyber-shot DSC-V3—Sony Corporation
Cyber-shot DSC-T3—Sony Corporation
DiMAGE G600—Konica Minolta Photo Imaging, Inc.
DCR-HC40—Sony Corporation
Hutchison Whampoa "3": 3G Mobile Videophone e228—
    NEC Corporation
SCH-V450—Samsung Electronics, Ltd.
DHG-HDD250—Sony Corporation
HD1000—Roku LLC
HDTV LCD projection televisions: LS47P1—Epson American, Inc.
HD2965FZD—SVA (Group)
Wide-screen "FD Trinitron WEGA": KD-34XBR960—
    Sony Electronics, Inc.
Transcend Digital Album—TS25PSP20G/20Gb
HDPS-M1—Sony Corporation
Photo Vault: MCS1—Sony Electronics, Inc.
Digital Printer: DPP-EX50—Sony Corporation
Digital Voice Recorder: ICD-BM1—Sony Electronics, Inc.
Digital Photo Frame: PHD-A55—Sony Corporation
Disc Steno Copier: CP100—Apacer Technology, Inc.
DVD Creation Station 200—SCM Microsystems, Inc.
PDA: CLIE PEG-UX50—Sony Corporation
PDA: CLIE PEGA-VR100K video recorder accessory—
    Sony Corporation
Palm OS PDA: Acer n20w—Acer, Inc.
Computer: Endeavor NT300—EPSON DIRECT Corporation

were three strange pieces of information on the product and packaging that helped lead me to everything that I needed to know about this product.

First, there was an e-mail address: info@armstation.com. Second, there was a Web site similarly named: www.armstation.com. Finally, the name, A.R.M. Kit, was stamped into the plastic on the accessory's bright blue cover plate. A quick trip to the ARM Station Web site gave me everything that I needed.

> **CHEAT CODE:** Established in October 2001, Sony Ericsson is a fifty-fifty joint venture of Sony Corporation and Ericsson AB.

Founded in 1999, ARM Station is a Hong Kong–based personal digital assistant (PDA), global positioning system (GPS), and mobile communications developer that designs and produces accessories and software for some major cell phone manufacturers. Organized into three divisions, ARM Station operates its PDA and mobile phone accessories line as A.R.M. Kit and its software branch as DailySoft. Its GPS-related products are distributed under the PalPal product line.

According to the Web site, ARM Station established business relationships with several major mobile communications manufacturers. Described as partnerships, these ventures included: Hewlett-Packard HK SAR Limited, JapanASK Corporation, Motorola Asia, Nokia (H.K.) Limited, O2 (Online) Hong Kong Limited, and Sony Ericsson Mobile.

> **CHEAT CODE:** You might still be able to buy an A.R.M. Memory Stick Jacket from eXpansys (www.expansys-usa.com). Just look for item number 105166, the Memory Stick Expansion Jacket. The cost should be approximately $30.00.

Continuing my search of the ARM Station Web site finally located some basic information about this product. Ah, the real name for this curious adapter accessory is A.R.M. Kit Memory Stick Jacket for P800. Furthermore, this adapter replaces the battery cover on the P800. One of the other components that is supplied with the A.R.M. Kit Memory Stick Jacket for P800 is a thin, plastic, 10-conductor ribbon cable with a Memory Stick Duo adapter connector (Fig. 4-3). This cable/connector combination actually serves as the conduit and converter for enabling older Memory Stick media cards to be used with the P800 in lieu of the Memory Stick Duo.

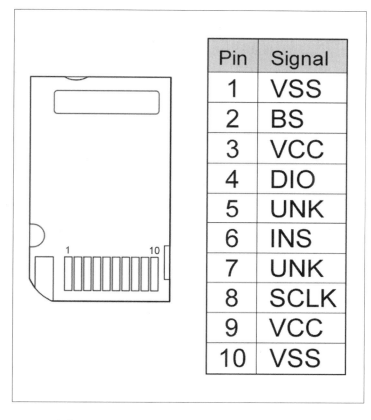

FIGURE 4-3
Memory Stick PRO Duo pin outs.

Here's how it works:

1. Prepare the P800 by removing the back cover and battery.

2. Carefully insert the Memory Stick Duo adapter plug in the P800 Memory Stick Duo slot.

3. The other end of the ribbon cable is attached to the inside Memory Stick connection port of the A.R.M. Kit battery cover replacement jacket.

4. Reinsert the battery into the P800, clean up your installation (i.e., remove any slack in the ribbon cable), and click the new A.R.M. Kit battery cover replacement jacket into place where the original battery cover was originally located.

HOW TO USE OLD MEMORY STICK MEDIA CARDS ON YOUR PSP

AS-1 All of the parts that you will need to use older Memory Stick media cards with the PSP. Maybe.

AS-2 If your A.R.M. Kit Memory Stick Jacket for P800 adapter won't fit in your PSP, peel the plastic off the underside of the Memory Stick Duo plug.

AS-3 Carefully, push the adapter plug into the PSP. This is a very snug fit.

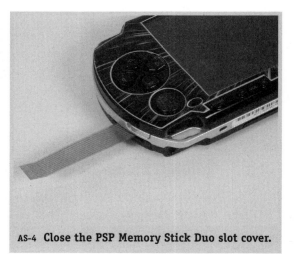

AS-4 Close the PSP Memory Stick Duo slot cover.

AS-5 Pull the ribbon cable connector lock out on the A.R.M. Kit Memory Stick Jacket.

AS-6 Insert the ribbon cable.

AS-7 All ready for older Memory Sticks. Maybe.

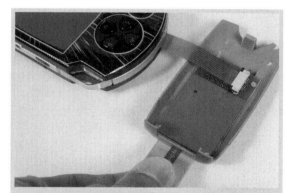
AS-8 I tested this A.R.M. Kit Memory Stick Jacket for P800 with Memory Stick Duo and Memory Stick PRO Duo media cards, and none of them worked. I used the Sony Memory Stick Duo Adaptor [sic] for these tests. I was unable to find any older Memory Stick media cards that would work, either.

ARMed But Not Dangerous

Now, you can start using older, bigger, and cheaper Memory Stick media cards with the P800. And should you wish to use Memory Stick Duo media cards with your phone, simply unplug the Memory Stick Duo adapter and remove the A.R.M. Kit Memory Stick Jacket from the P800. I'll bet that you can see where I'm going with this demonstration, can't you?

Well, don't believe it, although the A.R.M. Kit Memory Stick Jacket can be easily adapted to the PSP. There's more to recommend *against* using it rather than trying to recycle all of those older Memory Stick media cards that you have laying around.

> **CHEAT CODE:** ARM Station also makes an A.R.M. Kit– Screen Protector for the PSP. Known as the Crystal View Edition, this protector fits over the PSP's screen and guards it against smudges and scratches.

Yes, there is a fly in this ointment. During extensive testing of the A.R.M. Kit Memory Stick Jacket, I determined that it *might* work. Why am I not more conclusive about my claim that it will work? Because there are too many variables, and my tests were never repeatable nor conclusive. Case in point, Memory Stick Pro media cards wouldn't work. These test results were attempted on five different PSP systems, as well as on four Memory Stick Duo media cards ranging in size from 32 to 128 MB coupled to a Sony Memory Stick Duo Adaptor ([sic]; MSAC-M2).

Basically, the PSP would attempt to read the Memory Stick Duo, but couldn't recognize the media card's presence (i.e., the Memory Stick Duo access indicator kept flashing). Was the cause serial interface (original Memory Stick media cards) versus parallel interface (PSP)? Don't know. Would regular Memory Stick PRO media cards work? Don't know.

Therefore, I have illustrated my steps for you only as a demonstration of a Memory Stick interface that could provide some relief for readers who have a sizable investment in older Sony media cards. The bottom line is that I would not recommend this type of jury-rigged setup to any PSP user.

I Want My PSPTV

Hang on there, don't jump to any conclusions. This isn't what you might be thinking. Rather than listening to your music (MP3) without being able to watch your videos (MPEG4) simultaneously, how about making your own music videos (MP3 + MPEG4)? Forget playlists, album cover art, and music and videos sequestered away from each other in separate folders. Take the power of the PSP to its limit with your own music video—or, PSPTV.

Your Memory Stick PRO Duo is the ideal media for managing this feat. Just use your fav moviemaking app, a fistful of ripped MP3s, a stack or two of JPEG images, and a couple of hours worth of video footage, and, in less than 4 hours, you can have the ultimate PSPcast. See the next four pages for how.

HOW TO WATCH VIDEOS WHILE LISTENING TO MUSIC ON YOUR PSP

MS-1 Assemble the usual suspects that you prefer to use for movie making. I used iMovie by Apple Computer.

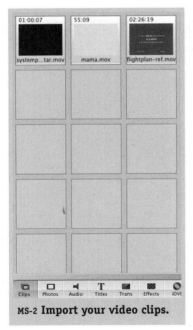

MS-2 Import your video clips.

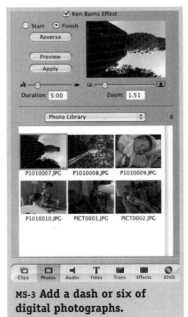

MS-3 Add a dash or six of digital photographs.

MS-4 Sprinkle liberally with MP3 music. Don't rely on music that you've purchased through Apple's iTunes Music Store. These tracks are protected with DRM software that cripples and limits your ability to import this music into projects like your music video. Instead rip your MP3s from CDs that you've purchased.

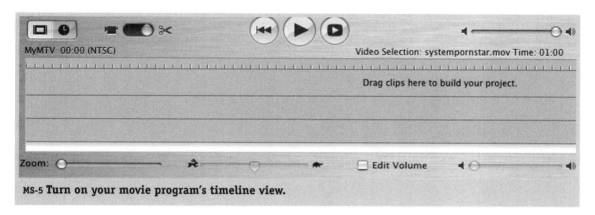

MS-5 **Turn on your movie program's timeline view.**

MS-6 **Begin importing both your video tracks and still images.**

MS-7 **Take some credit for your work. Create a meaningful title screen.**

MS-8 **Make a transition between your video and title tracks.**

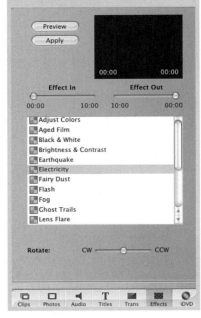

MS-9 **Still photos can be brought alive through a carefully selected and managed effect.**

MS-10 **Remove all of the audio portion from your video tracks.**

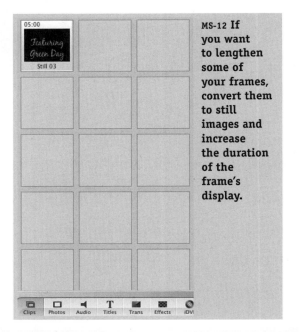

MS-11 **Drag your MP3 music into the appropriate frame sequence.**

MS-12 **If you want to lengthen some of your frames, convert them to still images and increase the duration of the frame's display.**

84

MS-13 **Now test your complete music video.**

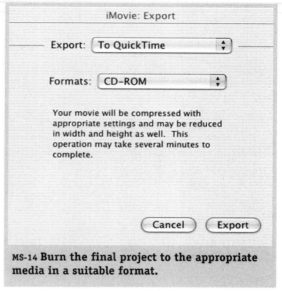
MS-14 **Burn the final project to the appropriate media in a suitable format.**

MS-15 **View the final project before you convert it to MPEG4.**

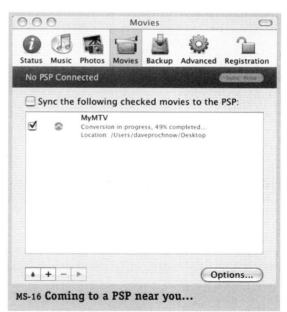
MS-16 **Coming to a PSP near you...**

CHAPTER 5

UMDware

Here we go again. Right? Here's a proprietary format that may, or may not, endure the test of time. And, in a digital lifestyle, time is measured in months, not years. Luckily for PSP owners, the UMD format doesn't appear to be a reprise of the Sony Betamax fiasco. Rather, the UMD has experienced two significant timeline events that would indicate a long and useful life span.

First, on June 24, 2005, the UMD format was officially approved as a "standard format" by Ecma International. On the heels of this format approval, SCEA released a statement that claimed that two UMD movie titles had surpassed 100,000 sales, *each*. Both *Resident Evil: Apocalypse* and *House of Flying Daggers* were cited by SCEA as having achieved sales exceeding 100,000 units. Comically, in the same statement, SCEA pointed out that *Spider-Man 2* "sales" were not included in its UMD movie sales figures. Duh; *Spider-Man 2* wasn't "sold," it was "included" inside the initial one million PSP Value Packs that were sold during the North American launch.

The second, and even bigger, reason that UMD *might* be here to stay is the widespread support that the format is receiving from various media publishers. Game creator names such as Activision, LucasArts, and Namco will help to ensure that the UMD format gets a solid foothold in the gaming arena. Likewise, major motion picture studios such as 20th Century Fox, New Line Home Entertainment, and Paramount Home Entertainment will help drive UMD sales. Furthermore, these studios are running two powerful marketing promotions for each UMD release that makes this new diminutive format seem more legitimate: UMD movies are released simultaneously with the ubiquitous DVD format movies, and both the UMD and the DVD releases are priced similarly.

Oddly enough, it isn't really a question about whether or not the UMD format is here to stay, rather it's a question of where *do* the UMD movies and games stay. In other words, where do you look for a UMD title? OK, the PSP UMD games aren't that difficult to locate. Just enter your fav consumer electronics megastore and saunter back to the gaming area. There you should find the PSP UMD game treasure trove. What about the UMD movies? Where do you look for a newly released title? In the gaming area, in the DVD section, or somewhere in between?

> **CHEAT CODE:** How can you tell that the UMD format is a popular format? When porn movies appear. Leading the way in making the UMD format really popular is Japanese porno giant GLAY'z.

Even as late as September 2005, Target, Best Buy, Circuit City, Toys "R" Us, and Wal-Mart were each experimenting with different methods for merchandising this fledgling format. Understandably, DVDs command the larger market share of movie sales, and with the now dead VHS store shelf space being slurped up by DVD products, UMDs are left as orphans: crammed on shelves with UMD games, slotted into end-cap racks in the DVD section, and, regrettably, sequestered behind glass cabinets in a netherland floating between games, entertainment, and toys.

One group of PSP owners who haven't had any trouble locating UMD media is the gamer crowd. Shortly after the introduction of the UMD, gamers were clamoring for multiplayer games. Basically, this type of game play enables two or more players to participate in the same game. And, on the PSP, multiplayer games can be fast and furious. Thanks, in part, to the ad hoc network capability that is built into the PSP. If a UMD game is programmed to accept multiplayer participation, then the ad hoc network system is activated. The only catch is that every participant must own a copy of the same game. Unfortunately, owning two or more copies of *any* UMD game can be an extremely costly (ad)venture. Therefore, this benign little hack was developed for circumventing the need to own two or more copies of a multiplayer game.

Before I show you how to frag your buddy's brains out, let me address one slight problem with this hack. It doesn't work 100% of the time. Anytime the game needs to access the UMD media, game play will come to a screeching halt. Other than that little tiny annoyance, however, this hack is rock solid and a great way to test multiplayer gaming before you invest in two or more copies of the same UMD game title. So, make sure that you select a PSP game title that *will* work. For example, *Wipeout Pure* will not work in this UMD swap hack. But, *Archer Maclean's Mercury* will work. Therefore, I will use *Mercury* as the PSP game for demonstrating the UMD swap hack technique.

Now, let's lock and load (Fig. 5-1).

FIGURE 5-1
The UMD access door is hinged from the back.

STEP 1. Insert the *Mercury* UMD into the first PSP, load the game, and select "Two Player Game."

STEP 2. Enter a name for the first PSP. For example, "ONE" is perfect for this demonstration.

STEP 3. Select "Start a Network Game." After a brief connection test to the wireless network, the first PSP will wait with the message "Waiting for Player to Join Game."

STEP 4. Remove the UMD from the first PSP and select the "No" response to the question "Do you want to quit the game?"

STEP 5. Insert the UMD into the second PSP, load the game, and select "Two Player Game."

STEP 6. Enter a name for the second PSP, such as "TWO."

STEP 7. Select "Join a Network Game" and press "OK" for the selection "<ONE> 0% Completed."

STEP 8. Accept the player "<TWO>" on the first PSP and wait while both machines communicate with each other. The message "Connecting the two machines together" will appear.

> **CHEAT CODE:** In an attempt to thwart homebrew developers, SCEA began using the UMD media as a means for requiring mandatory PSP firmware updates. In other words, if you don't have the "latest and greatest" version of the PSP firmware, then you don't get to play the UMD game. Granted, the newest firmware is generally supplied with the UMD game, but the requirement for a system upgrade *after* you've purchased a game is, well, naughty.

STEP 9. When the message, "The host is choosing which levels to play. Please wait," eject the UMD from the second PSP and insert it into the first PSP.

STEP 10. Select the "No" response to the question "Do you want to quit the game?"

STEP 11. Accept the selected level on the first PSP by pressing "OK" and then select between a "Race," "Percentage," or "Task" and press "OK."

STEP 12. The first PSP will now display "Synchronizing the two machines." Additionally, you will be reminded that: "Press the select button to stop waiting for opponent and quit the multiplayer game." Don't do this step unless you want to stop the multiplayer game. Duh? Eject the UMD from the first PSP and insert it into the second PSP.

STEP 13. Select the "No" response to the question "Do you want to quit the game?" Now play.

It might take a couple of tries to get all of this UMD swapping to orchestrate properly. If you're a little too slow on the exchange you could see, "Connection has timed out," and you could be disconnected from the multiplayer, wireless network. If this happens, just reset to Step 1 and try again.

UMD Night at the Movies

Here is a partial list of movie titles that are (or will be) available in UMD format for the PSP.

13 Going on 30: Sony Pictures Entertainment
A Knight's Tale: Sony Pictures Entertainment
Ah! My Goddess: The Motion Picture: Geneon
Air Force One: Sony Pictures Entertainment
Akira: Geneon
American Pie: Universal Studios
Anacondas: The Hunt for the Blood Orchid: Sony Pictures Entertainment
Appleseed: Geneon
Are We There Yet?: Sony Pictures Entertainment
Armageddon: Buena Vista Home Entertainment
Assault on Precinct 13: Universal Studios
Bad Boys: Sony Pictures Entertainment
Batman: The Movie: Twentieth Century Fox Home Entertainment (20th Century Fox)
Be Cool: MGM Studios
Beauty Shop: MGM Studios
Big Daddy: Sony Pictures Entertainment
Black Hawk Down: Sony Pictures Entertainment
Blade: New Line Home Entertainment
Blood: The Last Vampire: Anchor Bay Entertainment
Bobbit's Basics to Boogie Boogeyman: Sony Pictures Entertainment
The Bourne Identity: Universal Studios
Bulletproof Monk: MGM Studios
The Butterfly Effect: New Line Home Entertainment
Cabin Fever: Lions Gate Entertainment
Chappelle's Show: Volume 1: Paramount Home Entertainment
Charlie's Angels: Sony Pictures Entertainment
The Chronicles of Riddick: Unrated Director's Cut: Universal Studios
Coach Carter: Paramount Home Entertainment
Cowboy Bebop: 1st Session: Bandai Entertainment
Coyote Ugly: Unrated Extended Cut: Buena Vista Home Entertainment
Cursed: Buena Vista Home Entertainment
Daddy Day Care: Sony Pictures Entertainment
Dave Chappelle: For What It's Worth: Sony Online Entertainment
Dawn of the Dead: Unrated Director's Cut: Universal Studios
The Devil's Rejects: Lions Gate Entertainment

Dodgeball: A True Underdog Story: Twentieth Century Fox Home Entertainment (20th Century Fox)
Elf: New Line Home Entertainment
Endless Summer: Image Entertainment
Erotic Terrorist Beautiful Body NOA: GLAY'z
Escaflowne: The Movie: Bandai Entertainment [Video]
Eureka Seven—Volume 1: Bandai Visual
Evil Dead: Anchor Bay Entertainment
Family Guy Presents Stewie Griffin—The Untold Story: Twentieth Century Fox Home Entertainment (20th Century Fox)
The Fast and the Furious: Universal Studios
Freddy vs. Jason: New Line Home Entertainment
Friday Night Lights from Dusk till Dawn: Buena Vista Home Entertainment
Gankutsuou: The Count of Monte Cristo UMD 1: Geneon
Ghost in the Shell: Manga Entertainment
Ghostbusters: Sony Pictures Entertainment
Godsmack Live: Image Entertainment
Goku Hong—Riko Tachibana: GLAY'z
Gone in 60 Seconds: Unrated Director's Cut: Buena Vista Home Entertainment
Gundam Wing: Endless Waltz: Bandai Entertainment
The Grudge: Sony Pictures Entertainment
Gungrave Episodes 1 & 2: Geneon
h.m.p. Countdown 2005: h.m.p.
Halloween: Anchor Bay Entertainment
Hellboy: Sony Pictures Entertainment
Hellsing Episodes 1 & 2: Geneon
Hero: Buena Vista Home Entertainment
High Grade Class First Soap Lady—Anna Kaneshiro: GLAY'z
High Tension: Lions Gate Entertainment
Hikaru Hoshino BEST: h.m.p.
The Hillz: Image Entertainment
Hitch: Sony Pictures Entertainment
The Hitchhiker's Guide to the Galaxy: Buena Vista Games
Hollow Man: Sony Pictures Entertainment
Hoshi No Koe: The Voices of a Distant Star: CW Films
Hostage: Buena Vista Home Entertainment
House of the Dead: Lions Gate Entertainment
The House of Flying Daggers: Sony Pictures Entertainment
I, Robot: Twentieth Century Fox Home Entertainment (20th Century Fox)
The Incredibles: Buena Vista Home Entertainment

The Italian Job: Paramount Home Entertainment
Jamie Foxx: I Might Need Security: Image Entertainment
Kill Bill Volume 1: Buena Vista Home Entertainment
Kill Bill Volume 2: Buena Vista Home Entertainment
King Arthur: Extended Unrated Director's Cut: Buena Vista Home Entertainment
Lolicon Pick-Up 5: GLAY'z
The Magic of Flight: Image Entertainment
Man of the House: Sony Pictures Entertainment
Napoleon Dynamite: Twentieth Century Fox Home Entertainment (20th Century Fox)
Naruto: The Movie: Aniplex
National Lampoon Presents: The Best of the Romp Volume 1: Genius Products
National Lampoon's Van Wilder: Lions Gate Entertainment
National Treasure: Buena Vista Home Entertainment
Ninja Scroll: Manga Entertainment
Not Another Teen Movie: Sony Pictures Entertainment
The Nurse of a Big Breast—Mitsu Amai: GLAY'z
Once Upon a Time in Mexico: Sony Pictures Entertainment
The One: Sony Pictures Entertainment
Open Water: Lions Gate Entertainment
The Palace of a Virgin—Hikaru Coto: h.m.p.
Paranoia Agent—Volume 1: Geneon
Pinmen: Sony Pictures Entertainment
Pirates of the Caribbean: The Curse of the Black Pearl: Buena Vista Home Entertainment
Predator: Twentieth Century Fox Home Entertainment (20th Century Fox)
The Punisher: Lions Gate Entertainment
R.O.D. the TV—Volume 1: Geneon
Rambo: First Blood: Lions Gate Entertainment
Reign of Fire: Buena Vista Home Entertainment
Ren and Stimpy Volume 1-1/2: Paramount Home Entertainment
Resident Evil 2: Apocalypse: Sony Pictures Entertainment
Resident Evil: The Movie: Capcom
Richard Pryor: Live on Sunset Strip: Sony Pictures Entertainment
The Rock: Buena Vista Home Entertainment
The Rundown: Universal Studios
S.W.A.T.: Sony Pictures Entertainment
Sahara: Paramount Home Entertainment
Samurai Champloo Episodes 1 & 2: Geneon
Samurai Champloo Episodes 3 & 4: Geneon

Samurai Champloo Episodes 5 & 6: Geneon
Saw: Lions Gate Entertainment
Saw 2: Lions Gate Entertainment
Scryed—Volume 1: Bandai Entertainment
Secondhand Lions: New Line Home Entertainment
Shaolin Soccer: Buena Vista Home Entertainment
Short Circuit: Image Entertainment
Sin City: Buena Vista Home Entertainment
Snatch: Sony Pictures Entertainment
South Park: Volume 1: Paramount Home Entertainment
Spider-Man 2: Sony Pictures Entertainment
Spider-Man: The New Animated Series—Volume 1: The Mutant Menace: Sony Pictures Entertainment
SpongeBob SquarePants: The Movie: Paramount Home Entertainment
Stargate: Lions Gate Entertainment
Stargate: Atlantis Vol. 1: MGM Studios
Starship Troopers: Sony Pictures Entertainment
Steamboy: Sony Pictures Entertainment
Step Into Liquid: Lions Gate Entertainment
Super Troopers: Twentieth Century Fox Home Entertainment (20th Century Fox)
Team America: World Police: Paramount Home Entertainment
Tenjho Tenge—Round 1: Geneon
Terminator 2: Judgment Day: Lions Gate Entertainment
The Texas Chainsaw Massacre (2003): New Line Home Entertainment
Thumb War/Thumbtanic/Time Bandits: Anchor Bay Entertainment
Total Recall: Lions Gate Entertainment
Trigun—Volume 1: Geneon
Tron: Buena Vista Home Entertainment
Tupac Shakur: Live at the House of Blues: Image Entertainment
Tupac Shakur: Thug Angel: Image Entertainment
Universal Soldier: Lions Gate Entertainment
Van Helsing: Universal Studios
Viva La Bam: Volume 1: Paramount Home Entertainment
Waiting: Lions Gate Entertainment
Without a Paddle: Paramount Home Entertainment
XXX: Sony Pictures Entertainment
XXX: State of the Union: Sony Pictures Entertainment
You Got Served: Sony Pictures Entertainment
Young Guns: Lions Gate Entertainment
YUKI—Yuki Video: Epic Records

It Sounds Like...

Do you find the sound output from the built-in PSP speakers to be less than breathtaking? Likewise, aren't those white plastic earbud headphones more suited to being stuck anyplace else, other than your ears? Well, the answer is to amplify your PSP sound output. There are several great commercial options for answering this mail; alternatively, you can roll your own. If you're feeling frisky with the soldering iron, you can quickly, easily, and inexpensively build your own stereo 1-watt (W) PSP amplifier.

> **CHEAT CODE:** After the release of PSP Firmware 1.5.2, UMD media had a new member in the family—music.

I wanted to add some panache to my amp project, so I embedded the parts inside a WowWee Ltd. Robosapien. This is very similar to an iPod shuffle project that I featured in my book *The Official Robosapien Hacker's Guide* (McGraw-Hill, 2006).

Hack Highlights

Here's what you'll need to build your own 1-W stereo amplifier:
- (2) LM386 ICs
- (2) 10K potentiometers
- (2) 220-µF electrolytic capacitors
- (2) 8-ohm speakers
- (1) 24-inch, 3.5-mm stereo phone plugs cable
- (1) 3.5-mm stereo jack
- (1) 9-V battery
- (1) great looking case—such as a WowWee Ltd. Robosapien

A Hack in Stereo

STEP 1. This hack requires that you build a stereo amp circuit and install it plus two 8-ohm speakers inside the Robosapien (RS) chest cavity. Make sure that you install your 3.5-mm stereo jack near the neckline seam between the front and back chest plates. This black-painted plastic is perfect for hiding the jack.

HOW TO BUILD AN EXTERNAL STEREO AMP FOR YOUR PSP

AS-1 WowWee Ltd. Robosapien is an ideal PSP stereo amp enclosure.

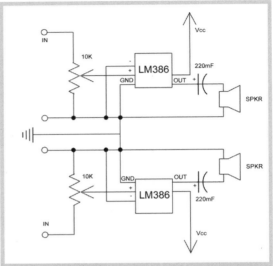

AS-3 Schematic diagram for a stereo amp circuit using two LM386 ICs.

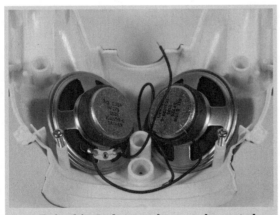

AS-2 Twin thin 8-ohm speakers can be nested nicely inside the front chest body plate of Robosapien.

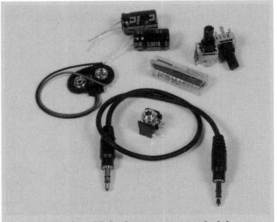

AS-4 Just a handful of parts are needed for building a stereo amp circuit.

HOW TO ADD STEREO THEATER SOUND TO YOUR PSP

TS-1 **Logic3 Sound System is theater sound for your PSP.**

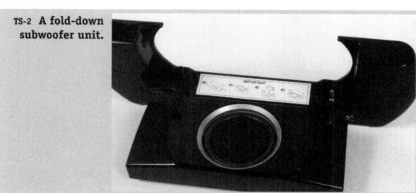

TS-2 **A fold-down subwoofer unit.**

TS-3 **Plug the PSP into the Logic3 Sound System making contact with the power connector and the line in connector.**

TS-4 Dynamic tweeter speakers drive the PSP sound output at a comfortable level.

TS-5 Battery operation enables the Logic3 Sound System to be placed anywhere in the home for portable UMD movie watching with theater-like surround sound.

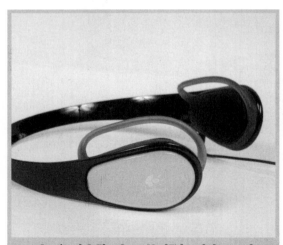

TS-6 Logitech® PlayGear Mod™ headphones for the PSP.

TS-7 Plug the PlayGear Mod headphones directly into the PSP.

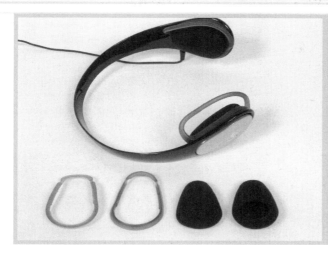

TS-8 A behind-the-head design and various comfort rings make this set of headphones perfect for any ear.

TS-9 Now two can listen to one, one PSP that is, with the Logitech PlayGear Share™ stereo adapter.

TS-10 Plug two stereo headphones into the PlayGear Share adapter.

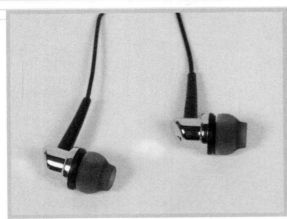

TS-11 Did you throw away the supplied Sony earbud headphones that came with your PSP? Good. Now get a great pair of ear bud headphones. The Logitech PlayGear Stealth™ headphones are comfortable and powerful rendering the stereo output in a realistic surround-sound experience.

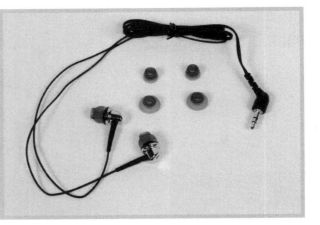

TS-12 There are three sizes of comfort ear gel plugs for the PlayGear Stealth so that no ear will be left behind.

TS-13 Logitech PlayGear Amp™ takes stereo sound on the road.

TS-14 Connect the PlayGear Amp to your PSP.

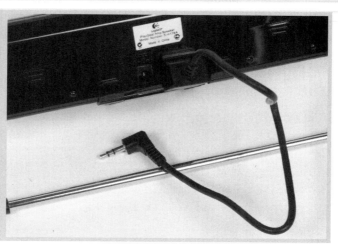

TS-15 Rotate the speakers into an upright position and attach the PSP cradle.

TS-16 Set the PSP volume to at least 50%, then plug in, turn on, and crank it up.

STEP 2. If you're able to locate some thin 8-ohm speakers, they should fit easily inside the front chest plate. Alternatively, you could mount one speaker in the front plate and the other speaker inside the back plate. This would give you better stereo sound, but the back plate is a tight fit, and the speaker would have to be insulated quite a bit around the main circuit board.

STEP 3. Wire your stereo amp circuit. Connect the twin speakers to this circuit, connect the 3.5-mm jack and battery, and plug in your PSP. Now adjust the 10K potentiometers for the best sound. If everything is a go, it's time for the show. Put the RS back together, attach the PSP, and program the RS to follow you around the house while mixing your favorite tunes, displaying a slideshow, or projecting your fav flick.

It's Game Time

This is a partial list of the UMD games (and distributors) that are currently available (or soon to be released) for the PSP.

50 Cent: Bulletproof: Vivendi Universal Games
Advent Shadow: Majesco
Ape Escape: On the Loose: Sony Computer Entertainment
Armored Core: Formula Front: FromSoftware
Astonishia Story: Sonnori
ATV Offroad Fury: Blazin' Trails: Sony Computer Entertainment
Battlefield 2: Modern Combat: Electronic Arts
Black & White Creatures: Majesco
BloodRayne: Majesco
Breath of Fire III: Capcom
Burnout Legends: Electronic Arts
Coded Arms: Konami
Command & Conquer: Electronic Arts
The Con: Sony Computer Entertainment
Crisis Core: Final Fantasy VII: Square Enix
Darkstalkers Chronicle: The Chaos Tower: Capcom
Daxter: Sony Computer Entertainment
Dead to Rights: Reckoning: Namco

Death Jr.: Konami
Devil May Cry Series: Capcom
Dragon Ball Z: Atari
Dynasty Warriors: KOEI
FIFA Soccer: Electronic Arts
Gagharv Trilogy 2: Bandai
Generation of Chaos: NIS America
Ghost in the Shell: Stand Alone Complex: Bandai
Ghost Rider: Majesco
The Godfather: Electronic Arts
Gran Turismo 4: Sony Computer Entertainment
Grand Theft Auto: Liberty City Stories: Rockstar Games
Gretzky NHL: Sony Computer Entertainment
GripShift: Sony Online Entertainment
Guilty Gear Judgment: Majesco
Gun: Activision
Harry Potter and the Goblet of Fire: Electronic Arts
Hot Shots Golf: Open Tee: Sony Computer Entertainment
The Incredibles 2: THQ
Infected: Majesco
Lemmings: Sony Computer Entertainment
Lumines: Ubisoft
Madden NFL 06: Electronic Arts
Marvel Nemesis: Rise of the Imperfects: Electronic Arts
MediEvil Resurrection: Sony Computer Entertainment
Mercury: Ignition Entertainment
Metal Gear Acid: Konami
Metal Gear Acid 2: Konami
Midnight Club 3: DUB Edition: Rockstar Games
MLB: Sony Computer Entertainment
Monster Hunter: Capcom
Mortal Kombat: Deception: Midway Games
MVP Baseball: Electronic Arts
Namco Museum Battle Collection: Namco
NBA: Sony Computer Entertainment
NBA Live 06: Electronic Arts
NBA Street Showdown: Electronic Arts
Need for Speed Most Wanted: Electronic Arts

Need for Speed Underground Rivals: Electronic Arts
NFL Street 2: Unleashed: Electronic Arts
Pac-Man World 3: Namco
PoPoLoCrois: Sony Computer Entertainment
Prince of Persia: Warrior Within: Ubisoft
Pursuit Force: Sony Computer Entertainment
Rengoku: The Tower of Purgatory: Konami
Ridge Racer: Namco
The Sims 2: Electronic Arts
Smart Bomb: Eidos Interactive
SOCOM: U.S. Navy SEALs Fireteam Bravo: Sony Computer Entertainment
Spider-Man 2: Activision
SSX on Tour: Electronic Arts
Stacked: Myelin Media
Star Wars Battlefront II: LucasArts
Star Wars Episode III: Revenge of the Sith: Ubisoft
Street Fighter Alpha: Capcom
Tales of Eternia: Namco
Tenchi no Mon: Sony Computer Entertainment
Tenchu: Shinobi Taizen: FromSoftware
Theseis: track7games
Tiger Woods PGA Tour: Electronic Arts
Tiger Woods PGA Tour 06: Electronic Arts
Tom Clancy's Splinter Cell 4: Ubisoft
Tomb Raider: Legend: Eidos Interactive
Tony Hawk's Underground 2 Remix: Activision
Twisted Metal: Head-On: Sony Computer Entertainment
Untold Legends 2: Sony Online Entertainment
Untold Legends: Brotherhood of the Blade: Sony Online Entertainment
Viewtiful Joe: Battle Carnival: Capcom
Virtua Tennis World Tour: Sega
WipEout Pure: Sony Computer Entertainment
Worms: THQ
X-Men Legends II: Rise of Apocalypse: Activision
Ys: The Ark of Napishtim: Konami

CHAPTER 6

The Modder of All PSPs

Have you seen any of the innovative uses of electroluminescent (EL) wiring that are starting to appear in automobiles, computers, and clothing? The flexible, low-heat, colorful wire is making significant inroads beyond the neon glow that typically illuminates gaming LAN parties (Fig. 6-1).

Just what is EL wiring? Those in the know call it EL wire. It also goes by the names of EL fiber, EL cable, and the popular "glow wire." I still prefer EL wire, but you make your own call. No matter what you call it, EL wire is a great product for modding virtually anything.

FIGURE 6-1
This baby's crying to be modded.

HOW TO MOD YOUR PSP WITH EL WIRE

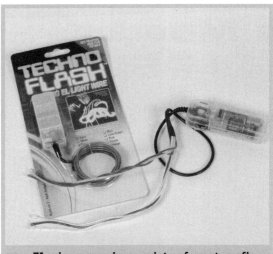

ES-1 EL wire comes in a variety of great configurations that are ideal for modding your PSP.

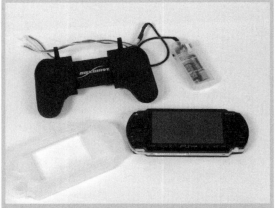

ES-2 Boxwave Corporation makes two great products that will make your EL wire modding easy: the PSP ActionGrip for holding the EL wire battery pack and the FlexiSkin for attaching the EL wire to the PSP.

ES-3 The back of the PSP is ideal for wrapping EL wire into some fun configurations.

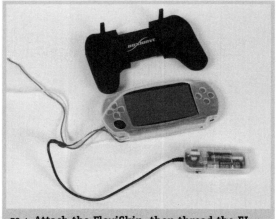

ES-4 Attach the FlexiSkin, then thread the EL wire around the PSP.

ES-5 Fix the wrapped PSP/FlexiSkin sandwich in the PSP ActionGrip.

ES-6 Use black cable ties for holding the EL wire battery pack firmly against the PSP ActionGrip.

ES-7 A tricked-out PSP and it can blink, too.

Wrap it around fabric, metal, wood, plastic, or baby sisters (that's a no-no), EL wire will not short, overheat, or discharge when in contact with these materials. Housed inside a plastic sheath, EL wire is actually a flexible copper wire that is coated with phosphorous. When low voltage is applied to this combination, the phosphorous glows. To supply the voltage to the copper wire, two transmitter wires are wrapped around the length of the EL wire. The external plastic sheath serves as a container for holding everything together and making EL wire a spiffy product that anyone can use for any purpose.

> **CHEAT CODE:** EL wire will gradually fade over time, with a typical life span measured in hours of use.

Powering the EL wire is accomplished by an inverter. The inverter generates a frequency that excites the phosphorous and makes a distinct colored glow at a specific intensity level. Most inverters operate in the 9- to 12-V range and typically require an AC power source. There are some terrific low-power alternatives, however. These EL wires can accomplish their distinctive glow with as little power as that provided by two AA-size batteries. This low-power option comes at a price—reduced length. Although their bigger AC brethren can drive 10 to 120 feet of EL wire, the pint-sized battery variety is limited to less than 3 feet. Luckily, when you're talking about modding a PSP, less than 3 feet of EL wire is a good thing.

There are some exciting "tricks of the trade" for making EL wire your own custom expression. Before you attempt to "customize" your EL wire, let me say two things. First, most EL wiring mods can be done with stock wire. You don't have to do any splicing or soldering, just plug your batteries in and create. Second, this type of customizing work requires some special tools (e.g., wire cutters, wire stripper, and soldering iron) and a basic understanding of electrical circuits. If you feel uncomfortable with these requirements or don't have the

The EL Wire You Say

Here are some sources for buying electroluminescent wiring:

AS&C CooLight—www.coolight.com

Beal Systems GLOWIRE—www.glowire.com

Elwirecheap.com—www.elwirecheap.com

Torche ELWIRE—www.elwire.com

necessary specialized talents, just use your EL wire "straight." If you'd rather "roll you own," read on.

Most EL wire is sold in single-color lengths or spools. Although this "monoglow" is great for a long, single-color run of wiring, you might want to mix your own palette. In this case, all you have to do is splice various colors of EL wire together. Just cut the EL wire, strip the ends, solder the two wires together, and seal the joint with some epoxy sealant.

CHEAT CODE: EL wire draws less than 1 A of current for approximately 100 feet in length.

Another creative trick with EL wire is to splice a regular non-EL copper wire to the inverter as an extender. This extender won't glow, but it will help in adding some extra space or separation between the inverter and the EL wire.

Considering all of the benefits that can be derived from EL wire, it's a wonder that you don't see it everywhere...even on your PSP.

Glow Little PSP, Glimmer, Glimmer

There's no denying that the "stock," out-of-the-box PSP is a marvelous demonstration of a near-perfect design. What are its failings? Well, for one the L (left) and R (right) buttons don't light up. In fact, none of the buttons light up, which makes using the PSP in the dark a very difficult proposition (Fig. 6-2).

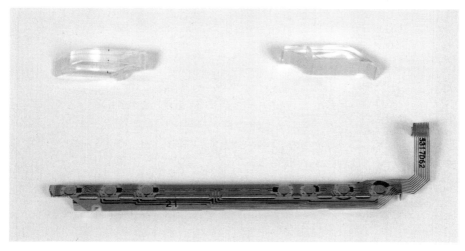

FIGURE 6-2
The L and R buttons and the bottom button bar are great locations for experimenting with some PSP mods.

HOW TO MOD YOUR PSP WITH LED BUTTONS

LS-1 You can pull your LED power requirements from a couple of different sources. If you elect to use the charger's lines, then your LED mod will only shine during your recharging activity.

LS-2 A better source for power would be from the main circuit board's V_{DD} pad.

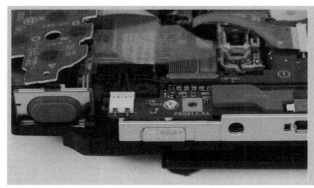

LS-3 This location is prime real estate for holding the LED mod.

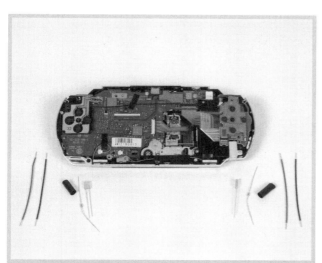

LS-4 All you need for adding a couple of green LED glow bugs to your PSP is a few discrete components.

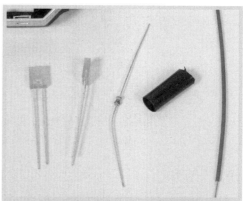

LS-5 Flat LEDs work best (e.g., Digi-Key part number 67-1046-ND). You will also need one 1.0K-ohm, $1/8$-W resistor for each LED. A short piece of heat shrink tubing and some spare hookup wire completes the parts list for this LED mod.

LS-6 Squirreling this mod into the right location can be a real exercise in circuitry contortion.

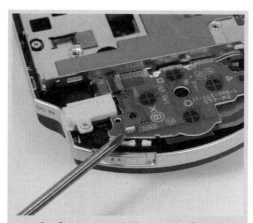

LS-7 Surface-mount LEDs are used on the PSP circuit boards for a similar effect.

CHAPTER 7

Get Yer Game Face On

Every time I play a game on my PSP, even a classic-rooted game such as *Wipeout Pure*, I marvel at the crisp vivid graphics, the snappy game play, and the über-portable form factor. As remarkable as the PSP is at gaming, though, this isn't the first time that a handheld game device has made a splash in the marketplace (Fig. 7-1).

Just a quick thumbing-through of *Phoenix: The Rise and Fall of Videogames* by Leonard Herman will help reveal the entire history of portable game machines. Likewise, studying this lineage will help to illustrate how the PSP took handheld game devices to the next level.

FIGURE 7-1
This thing is built for games.

There are six milestone products in the history of personal, portable gaming machines: Mattel Electronics LED, Nintendo Game Boy, Atari Lynx, Nintendo Game Boy Advance (GBA), Nintendo DS, and Sony PSP. Although each product made a contribution to the introduction/advancement of handheld game devices, they were also bellwethers of a unique platform that quickly became a dominant player in the video game industry.

MATTEL ELECTRONICS LED: Introduced in 1976, these low-tech, primarily red LED-driven games marked the first major inroad by a portable gaming device into the mainstream consumer marketplace. Remember when RadioShack stores would sell these products around Christmas time?

NINTENDO GAME BOY: The original incarnation of the Game Boy franchise was begun in 1989. Since then, the Game Boy lineup has become the most financially successful portable gaming device in history.

ATARI LYNX: "So close, yet not even close enough" became the epitaph of the Lynx. Released in 1989, yes, in direct challenge to the Game Boy steamroller, the Lynx's color display and unusual wide landscape form factor weren't enough to stem the tide of the Game Boy tidal wave.

NINTENDO GAME BOY ADVANCE: Twelve years later and still going strong, the Game Boy line received a much needed makeover in 2001 with the GBA. Now in color and sporting a wide landscape form factor (thank you, Lynx), the GBA became the new millennium's de facto portable gaming device. Not content to leave well enough alone, Nintendo updated the GBA to the klutzy clamshell Game Boy Advance SP.

> **CHEAT CODE:** In June 2005, Nikkei Business Publishing reported that Nintendo had set a manufacturing goal of 20 million Nintendo DS units for 2005.

NINTENDO DS: Responding to Sony's impending entry into the portable gaming market, Nintendo released the Nintendo DS in 2004. Considered to be the Game Boy Advance SP on steroids, the DS is a dual-screen, flip-up, wide landscape form factor with a curious touch-sensitive screen for stylus control and data entry. Although a remarkable design innovation that is still lacking in a UMD-like media library (Fig. 7-2), the DS is, nonetheless, giving the PSP some strong competition.

FIGURE 7-2
Armored Core®: Formula Front— Special Edition combines both action and strategy for entertaining game play from UMD media. (Photograph courtesy of Agetec, Inc.)

SONY PSP: Hail to the king, baby. The handheld gaming industry changed in 2004 with the release of the much-anticipated Sony PSP.

Hack Your Own GBAmp

After looking down portable game's memory lane, it's tough not to focus on the 2001 GBA. This is a potent gaming device that could be still found as recently as Christmas 2005 at a fire-sale price of less than $50. Given this low cost and its large user base, it stands to reason that if videos could be displayed on the GBA, this would be an interesting option to the PSP. You know, kinda like a poor boy's PSP. Well, this type of hack *is* possible, and all you need is the Flash Linker & Card Set.

CHEAT CODE: According to Nikkei Business Publishing, in mid-2005 Sony announced to its PSP suppliers that its yearly production forecast had been reduced from 18 million PSPs for 2005 to 12 million units.

I present to you the GBA media player or GBAmp.

To hack your GBA into a GBAmp, there is one little piece of magic that you will need: the Flash Linker & Card Set (http://gameboy-advance.net/flash_card/gba_X-ROM.htm). There has been a tremendous price drop in this product. I was able to purchase one for less than $70 from Easybuy2000.com.

HOW TO TURN YOUR GBA INTO A GBA MEDIA PLAYER

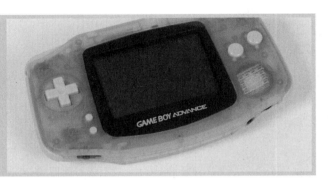

GS-1 Teach this old dog some new tricks. Convert a 2000-era GBA into a 2005-era media player.

GS-2 All of your old game cartridges can be easily backed up to your PC hard disk drive. These backed up ROM images can then be loaded onto a larger capacity 512-Mb Flash ROM cartridge.

GS-3 A 512-Mb Flash ROM cartridge can hold almost 30 GBA-ROM images, as well as some music, images, videos, and even e-books.

GS-4 A Flash Linker connects the GBA to your PC USB port.

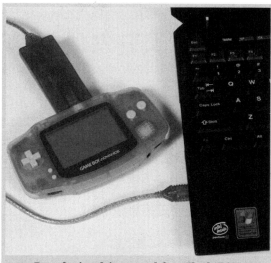

GS-5 Free device drivers and free Flash ROM burning software make the conversion of GBA to GBAmp painless and fun.

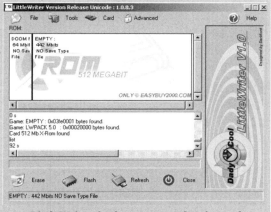

GS-6 LittleWriter is the program that is used for assembling your ROM images and flashing them onto your new 512-Mb Flash ROM cartridge.

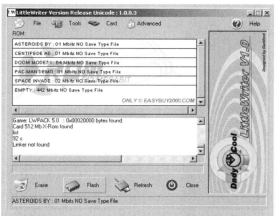

GS-7 A graph view in LittleWriter enables you to quickly survey the menu structure of the Flash ROM prior to flashing it.

The Flash Linker & Card Set consists of a 512-Mb GBA cartridge and a nifty USB-to-GBA interface. All of the software that supports this product can be downloaded from the Easybuy2000.com site: These are primarily LittleWriter and the X-ROM driver. You should download and install *both* of these programs *before* you connect the Flash Linker & Card Set to your Windows-based PC.

Once you're loaded, you're armed for complete GBA ROM cartridge copying to your PC, moving all of the desired ROM images onto the new Flash Linker & Card Set 512-Mb GBA cartridge (this cartridge can hold almost 30 ROM images), and plugging the new 512-Mb GBA cartridge into your GBA.

But that's hardly multimedia, is it? In addition to ROM image copying, the Flash Linker & Card Set is backed with some powerful programs for creating e-books, encoding video with the METEO codec, and assembling photo slideshows, all for playback off the 512-Mb GBA cartridge. You can even opt for a menu shell system for listing all of the stuff that is now crammed onto your GBA flash cartridge.

Is the GBAmp a threat to the PSP? Are you kidding? Although this hack is a super method for repurposing an outdated portable game machine, it doesn't hold a candle to the performance prowess packed into the PSP. For example, according to the GBA Web site, the best video quality that can be tweaked from this system is the following:

- Bitrate: 120 to 250 kbps
- Frame rate: 24 fps
- Prefilter: 15 bit/4 times

Once You Taste Homebrew, There's No Going Back

If you wished that there was some way that you could expand the PSP's game library past those titles on UMD (Fig. 7-3), then look no further than home-brew games. Homebrew is a four-letter word to Sony. Repeated firmware updates have been released in an attempt to thwart the efforts of the fledgling, yet highly ambitious, homebrew market. Leading the way in these homebrew efforts is the PSP-development team and its remarkable product KXploit.

KXploit (Direct Loader) version 1.50 is an exploit of the PSP firmware that enables users to load and play games and other applications from a single Mem-

FIGURE 7-3
Nintendo DS wishes that its games looked this good. *PoPoLoCrois* for the Sony PSP places you inside a Japanese anime series where you meet a lot of interesting characters. (Photograph courtesy of Agetec, Inc.)

ory Stick PRO Duo media card. The magazine *Make: Technology on Your Time* has been following the exploits of KXploit since its initial two–Memory Stick version. And it is this type of press coverage of this PSP firmware hack that has thankfully produced an air of legitimacy for the homebrew effort. So you can be confident that this is a powerful hack that works reliably with no ill effects on your PSP.

OK, now you know how to add homebrew games to your PSP; so what? Well, how about this terrific homebrew classic: *DOOM*? Yup, an innovative PSP hacker, known as lantus (www.lantus-x.com/PSP), ported one of the all-time-classic first-person shooter titles over to the PSP. Although the initial porting effort met with a tepid reception (e.g., the sound quality was admittedly crappy), lantus kept hacking and, in late July 2005, released a full-screen mode, stereo sound masterpiece. There are some quirks in installing and running the lantus PSP *DOOM* port that can cause even veteran PSP users to scream for help. So for all of you n00bs, see the next two pages for a step-by-step tutorial for enjoying the classic shareware *DOOM* on your handheld entertainment shotgun, err, system.

Another avenue for enjoyment with your homebrew PSP is in the emulation of other game systems. By using special emulators, the PSP can be tricked into mimicking a variety of other game systems. Systems such as Amiga, Neo GeoNeo Geo, and Sega can be resurrected from the dead game system grave-

HOW TO PLAY HOMEBREW GAMES ON YOUR PSP

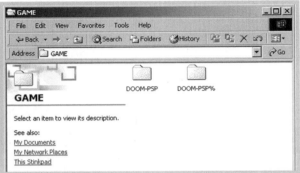

HBS-1 The homebrew PSP game version of *DOOM* by lantus consists of two folders that are saved inside the game folder on your Memory Stick PRO Duo media card.

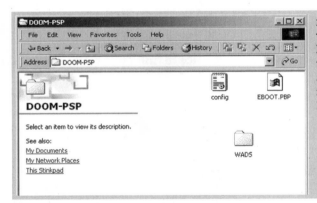

HBS-2 One folder titled DOOM-PSP holds the shareware DOOM WADS, a configuration file, and an EBOOT file.

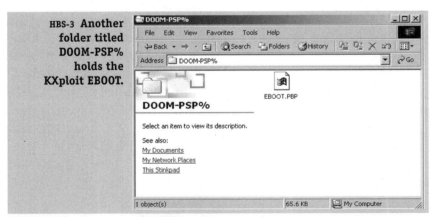

HBS-3 Another folder titled DOOM-PSP% holds the KXploit EBOOT.

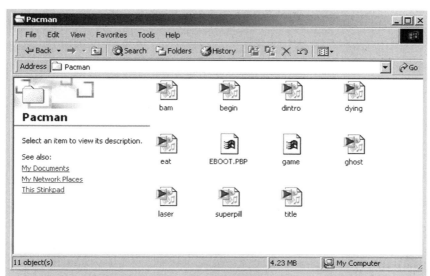

HBS-4 **Other homebrew games can safely coexist on the same media card. For example, Jim Peerless and Team Discordia have made a nifty Pacman clone: PSPacman for Firmware 1.50 PSP owners.**

HBS-5 **This is homebrew nirvana, baby—Firmware 1.50. Check your system settings before you attempt to use homebrew software that is built for a specific firmware version.**

> **EASTER EGG:** For the most part, Mac gamers have been left out in the cold when trying to run homebrew games on their PSPs. Programs such as KXploit and PSPE only run on Windows-based PCs. From the makers of iPSP, RnSK Softronics (http://ipsp.kaisakura.com/index.php), comes HomebrewPSP Converter for Mac OS X 10.4.x. After you attach your PSP to your Mac via a USB connection, just add your EBOOT.PBP PSP program files to HomebrewPSP Converter and click the "Transfer to PSP" button. This app will create the needed folder hierarchy and copy the program files to the PSP Memory Stick PRO Duo media card.

yard through the use of emulators. One place to look for these emulators is on the Content Holdings LLC Web site at www.pspupdates.com. It isn't as simple as loading a homebrew emulator into your PSP to start playing all of the classic Sega titles. You will need to include a particular game's ROM image file on the PSP along with the emulator. This process is very similar to the installation of *DOOM* homebrew on your PSP. So once you master the lantus *DOOM* port, then load up your Memory Stick PRO Duo with an emulator or three and make a portable dash into the past.

PSP Emulators Available at www.pspupdates.com

- Amiga 500
- Chip8
- Game Boy Color
- GBA
- Genesis/Megadrive
- Images
- MAME
- MSX
- Neo Geo
- NES
- PC Engine
- PC-9801
- Playstation
- SCUMM
- Sega Master System
- SNES
- Windows
- Wonderswan

Roll Your Own

Once you're bitten by the homebrew bug, you quickly realize that your PSP could be more than just a terrific UMD game and movie system; it could be made into *your* PSP. In other words, you could customize your PSP with your own brand of games.

For example, do you wish that you had a file manager for browsing the homebrew folder hierarchy of your PSP media card? So did Nevyn (Joachim Bengtsson) and the result was that he wrote Lowser for the PSP.

If only you could program, right? Lucky for you, you *don't* really need to know how to program for rolling your own PSP apps. Unlike other game systems, the PSP has a terrific scripting language that can be used for building anything from games to utilities. It's called Lua Player.

Lua Player is an exciting script-like programming environment that is derived from the Lua Programming Language (Fig. 7-4). Never heard of Lua? Jeez, you need to get out more. Lua is an extensible ANSI C-like interpreted language that was developed by a team at Tecgraf, the Computer Graphics Technology Group of the Pontifical Catholic University of Rio de Janeiro in Brazil.

FIGURE 7-4 **Lua Player is based on the Lua Programming Language.** (Logo courtesy of Tecgraf, the Computer Graphics Technology Group of the Pontifical Catholic University of Rio de Janeiro in Brazil)

Frank Buss and Joachim Bengtsson (the same person who created Lowser) took the Lua language and packaged it into a PSP scripting environment. The result is Lua Player. Lua Player is a free download and comes with several great sample apps: Lowser, music test, calculator, clock, Snake game, Hello World, fractals, and a text rotation demo.

> **CHEAT CODE:** Lua is moon in Portuguese.

To roll your own Lua homebrew code, all you need is a text editing app, such as the venerable BBEdit. There is no need for a compiler or an integrated development environment (IDE). Just write your script and copy it onto your PSP media card. That's it. You're now the captain of your PSP.

Finally, the script for Lua Player is relatively easy to understand and implement. Consider this extract from Buss's Lua Player tutorial's Hello World sample:

green = Color.new(0, 255, 0)

screen:print(200, 100, "Hello World!", green)

Please refer to the entire script sample on the Lua Player Web site. Now get out there and code the next great PSP game.

Why Don't You Just Act Like a PSP?

As the homebrew gaming effort grows, there might be times when you wish that you could test a game prior to loading into KXploit and copying it onto your PSP's Memory Stick PRO Duo. Well, you're in luck—that is, if you have a Windows-based PC. The PSP emulator for Windows (PSPE) is a simple, no-menu application shell that loads and executes PSP EBOOT.PBP files, or PSP program files. Just set up a folder hierarchy for holding your homebrew game program files, start PSPE, and play, err, test your game.

> **CHEAT CODE:** If you intend to enter homebrew game playing, then you will have to learn about RAR. RAR is a file compression algorithm developed by Alexander Roshal's RARLAB and is the most common file archive format used with homebrew games. You can decode RAR file archives with either WinRAR (www.rarlab.com/download.htm) or StuffIt Deluxe by Allume Systems (www.stuffit.com/mac/index.html).

CHAPTER 8

Don't Forget the Sunscreen

Unlike any other personal consumer electronics product, the PSP is crying out for protection. Just one scratch on the beautiful display, for example, and your movie-viewing experience immediately goes from theater-quality to bathroom-quality (Fig. 8-1). And, that isn't to say that you have to be a five-thumbed klutz to ruin the PSP's display, either. All you have to do is look at the poor-quality protective accessories that Sony packed inside each initial PSP Value Pack, and you will quickly see that a scratch is definitely in your PSP's future.

On paper, the PSP Value Pack accessories look impressive:

→ Wrist strap—ugh, a white leather strap; what was Sony thinking? And, what's going to happen to your PSP as you dangle it from your wrist? Don't use it. Attach it to your cell phone or some other utilitarian consumer electronics product.

→ Pouch—this thing is a breeding ground for hairs, fuzz, and scratches. Who would even consider sliding a PSP into a tight, synthetic pouch? Not me, and don't you do it, either.

→ Polishing cloth—the one redeeming accessory included with the PSP. But, it's so small that it's hard to hold effectively. Similarly, it can easily trap harmful debris in its fibers. If you use it, and you should, regularly,

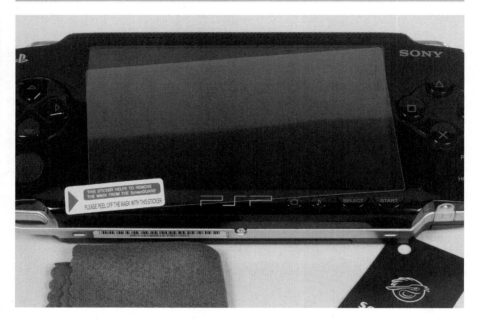

FIGURE 8-1
Pelican makes a number of products that help to protect your PSP.

make sure that you shake the cloth out periodically. Also, be sure to use a light touch when polishing and cleaning the PSP. Press too hard and you could leave a permanent smudge or streak. And, this is the type of streak that you won't be able to wash out in the laundry.

So, how can you protect your PSP without compromising its aesthetics? There are two options that you should consider: skin and case.

Get Skinned

In its simplest definition, a PSP skin is just that, a protective membrane that covers your PSP. Gamer Graffix™ takes this simple definition and creates a protective membrane that is a fashion statement—they call it "geek chic." Based on a flexible epoxy coating, Gamer Graffix Skinz are reusable, washable, and very personal. Don't be confused, these are *not* your grandfather's skins, which were actually stickers that gummed up your game machine and voided your factory warranty.

EASTER EGG: Five questions with Chris (Whitey) White, President/CEO of Gamer Graffix (www.gamergraffix.com):

1. DAVE PROCHNOW: What did you see in the handheld gaming device market that inspired you to get skinned?

CHRIS WHITE: All of the major device makers produce a very high quality product that really does look great. When we did our market research, we found only very sterile-looking cases retailing for 30 to 40 dollars. We knew we could do much better with the style, raise the bar on quality, and bring down the price to where it should be.

2. DP: Clearly the release of the Sony PSP has rekindled an interest in portable gaming. What did you do with your skin marketing to ensure that owners would want to get their PSPs skinned?

CW: Honestly? As with all new products, our Web site launched our Skinz 45 days prior to any stores receiving. It was such a smash hit right out of the box, all we really had to do was put it out there. The market responded by exceeding our expectations fivefold.

3. DP: Unlike other handheld gaming devices, Sony appears to be marketing its handheld toward a digital lifestyle rather than as a traditional gaming system. This type of concept has an emphasis on the PSP as a chic fashion accessory. How do skins fit into this digital lifestyle?

CW: We knew that the fashion aspect was one of the main reasons why our skins are so sought after. We give our fans an opportunity to dress, if you will, their systems to fit whatever mood they are feeling at that time. We are all about individuality!

4. DP: Have you performed any focus group tests on how a typical PSP owner would get skinned? Is there a special demographic that represents the ideal PSP owner for getting skinned?

CW: Gamer Graffix operates a Web site that gets millions of hits a month and more fan mail then we can count. We use this site as our test bed to determine the direction of our art team for future skin designs. We LUV our Web customers; they are truly some of the most informed consumers in the world! We really do read their mail, and we really do take their advice.

5. DP: In the upcoming year, what does Gamer Graffix have in store for PSP owners?

CW: What you have seen thus far from Gamer Graffix is just the tip of the iceberg. Our patent-pending process allows us to do more with our Skinz then any other known company in the world. I do not want to divulge too much, so let's just say we are sure every one will be blown away when we release Series 2 Skinz. We are also sure there is a huge market for reasonably priced custom Skinz...and that's all I am saying!

HOW TO SKIN YOUR PSP

GS-1 Gamer Graffix makes a Skinz for every season and most consumer electronics, too.

GS-2 Gamer Graffix PSP Skinz. And this beauty ain't just skin deep, either.

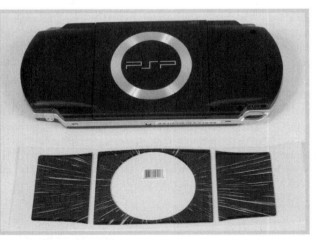

GS-3 Start with skinning the PSP back. There are three separate epoxy films for the back.

GS-4 Carefully apply the first film.

GS-5 I prefer to keep the Skinz backing paper attached as I roll the film off onto my PSP.

GS-6 There is one film for the PSP face.

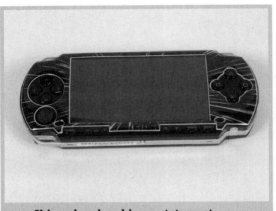

GS-7 **Skinned and making a statement. Unlike other game console skins, Skinz can be removed, discarded, and changed, keeping your PSP fresh and attractive.**

GS-8 **Skinned and not skinned.**

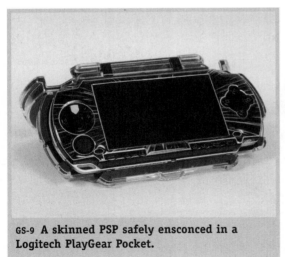

GS-9 **A skinned PSP safely ensconced in a Logitech PlayGear Pocket.**

GS-10 **A geek chic PSP fob.**

HOW TO WRAP YOUR PSP IN FINE FRENCH LEATHER

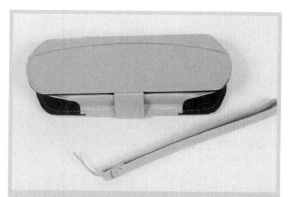

NS-1 What could be better for your PSP than some fine French leather holding it tighter than your fingers while flinging immune blood droplets in *Infected*? The Norêve PSP case is the Champagne of PSP cases.

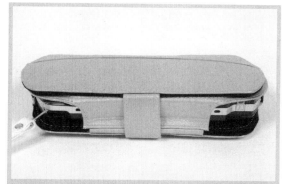

NS-2 There are ample openings in the Norêve PSP case for gaining access to all switches, plugs, and ports.

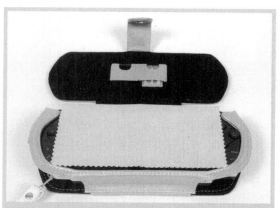

NS-3 I use the Sony-supplied polishing cloth for protecting the PSP LCD screen during storage.

NS-4 The Norêve PSP case has the ability to hold two Memory Stick PRO Duo media cards. Beware of those media cards when you close your case, however. I use my polishing cloth to protect the screen when the case is closed. I mean, who can turn down French leather?

Gamer Graffix has made some significant strides in the research, design, and packaging of PSP skins. Unlike other sticker-type skins, Gamer Graffix Skinz are packaged in an informative box that provides complete instructions for adding a skin to your PSP. Likewise, this intelligent packaging also supplies some care and feeding hints for keeping your PSP beauty more than skin deep.

Would You Like Some PSP with That Whine?

One scratch on your PSP and you'll end up crying like a little baby. To ensure that you have maximum protection for your PSP, you should consider a case for it. A real case. A real leather case. A real leather French case. Oui.

Norêve makes elegant leather cases for handheld and mobile devices. Located in Saint-Tropez, France, Norêve cases are well known for both their quality construction and their unique design ambiance. Ensconced in a supple and durable leather, portable electronics devices are not only protected inside Norêve cases, they are also enriched with a distinctive emblem that screams innovation.

> **CHEAT CODE:** Gamer Graffix also makes skins for Apple Computer iPod, iPod mini, Game Boy, Game Boy Advanced SP, Game Cube, Nintendo DS, PS2 (both original and the newer slim version), and XBox.

Right on the heels of the North American launch of the PSP, Norêve released their first case for this revolutionary handheld entertainment system. Matching the elegance in its portable cargo's design, the Norêve PSP case quickly flips open to grant complete access to every button, slot, and port. A soft velvet interior keeps dust and fuzz to a minimum, as well as helping to prevent scratches on the display. Some great storage for two Memory Stick PRO Duo media cards and one UMD movie or game is also, somehow, squeezed into the case. As easily as it opens, the Norêve PSP case snaps closed with a clever little magnetic closure (don't worry about magnetic interference, there is none). Available in either black or gray leather, the Norêve PSP case inspires cachet and renders PSP the de rigueur portable game machine.

Two recent additions to the growing Norêve product line are the following:

TRADITION LEATHER CASE FOR PSP—FOR WOMEN ONLY

- Elegant case in high-quality soft leather
- Access to basic functions (multiple openings on the leather)
- Snap button closure
- Two Memory Stick card slots
- One UMD slot
- Velvet interior lining
- Sold with leather wrist strap

SONY UMD AND MEMORY STICK KEEPER

- Elegant case in high-quality soft leather
- Eight Sony UMD slots
- Four Memory Stick slots
- Zipper closure

Don't Cry for Me PSP

If you own a consumer electronics product (e.g., cell phone, digital camera, MP3 player, notebook computer, PDA, BlackBerry, or PSP) and you don't know the name Vaja, then you must not care about fine leather cases that can protect your priceless gadget from the bumps and bangs of everyday life. Vaja leather products use the finest full-grain Argentine leather that is specially handcrafted by artisans who bring a unique luster and beauty to each case creating a handmade, custom-looking finish that you'd be hard pressed to find in any other case.

Oh, and the smell of a handmade Vaja Argentine leather case is divine. Your PSP will suddenly smell like an elegant electronics lifestyle fashion statement—just like it should.

> **CHEAT CODE:** Norêve cases are made from two types of leather: classic soft leather known as "Tradition" and a unique rigid leather known as "Ambition."

Finally, if you're looking to make a unique, personal statement about your PSP, then the Vaja online customization system is for you (www.vajachoice.com). At this Web site, you can select the exact design for your PSP i-volution Leather Suit case. Choose front and

HOW TO PUT ON A VAJA LEATHER SUIT

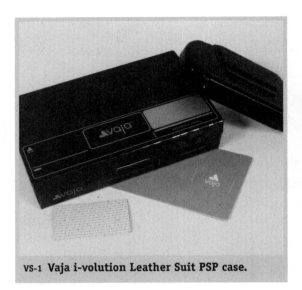

VS-1 **Vaja i-volution Leather Suit PSP case.**

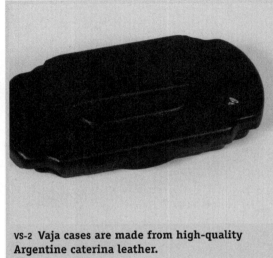

VS-2 **Vaja cases are made from high-quality Argentine caterina leather.**

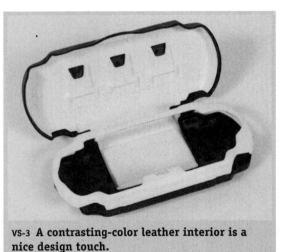

VS-3 **A contrasting-color leather interior is a nice design touch.**

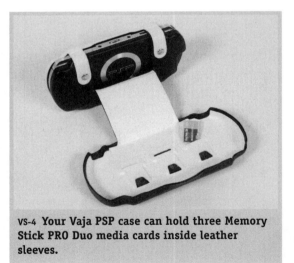

VS-4 **Your Vaja PSP case can hold three Memory Stick PRO Duo media cards inside leather sleeves.**

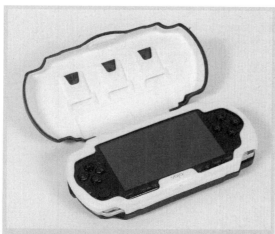

VS-5 **The top of the Vaja case folds out of the way when using the PSP.**

VS-6 **The Vaja case does *not* restrict access to the PSP UMD drive.**

VS-7 **Two simple straps hold your PSP inside the Vaja case.**

VS-8 **You can get a sense of the wonderfully molded leather of the Vaja i-volution Leather Suit PSP case in this close-up of the Memory Stick access port.**

back caterina leather color options, add an ultra belt clip, and print personalized text (up to three lines of text) and/or graphics on your case. The resulting case will take approximately 20 days to create. But, once you have your PSP safely tucked inside your custom Vaja case, you will have both the coolest game system on Earth *and* the coolest leather case ever designed—all in the palm of your hand.

The Case Case

Granted, a fine leather case might be beyond your means, but protecting your PSP investment is still a wise thing to do. There is a large assortment of cases, covers, and protectors that you can use to cradle your precious cargo. See the next three pages for the best ones that are currently available.

The Paper Case

One of the simpler, yet more visually exciting mods that you can make to your PSP is with the Logitech PlayGear Pocket™ case. An exciting feature about this case is the ability to remove the rubber liner and insert some custom artwork between the liner and the case. When the liner is reinserted, your custom artwork will scream your personality.

Although I have provided you with a great custom skin for your PlayGear Pocket, you might want to make your own. One source for artwork can be found on the Adobe Studio Exchange Web site (http://share.studio.adobe.com/default.asp). The shapes that you download from this Web site can then be saved in the Presets folder as Custom Shapes for use in Adobe Photoshop. Examples of the shapes that I used are: tiger images from Jewel's Knights by Brandy Wallen and dragon images from Wicked Dragons by Michael Norman. (See pp. 140 to 143.)

HOW TO SELECT A CASE FOR YOUR PSP

CS-1 Pelican Face Armor.

CS-2 You screw the Pelican Face Armor into those two screws holes located on the top side panel of your PSP. You wondered what those holes were for, didn't you?

CS-3 A hinge holds the Pelican Face Armor up and out of your way while you're playing a game or watching a movie. This screen protector also functions as a pretty good sunscreen while using the PSP outdoors.

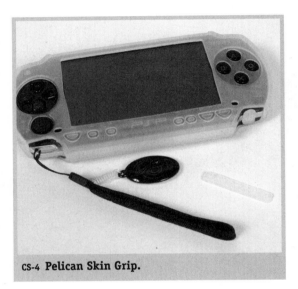

CS-4 Pelican Skin Grip.

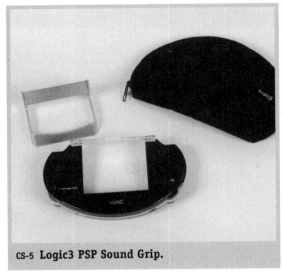

CS-5 Logic3 PSP Sound Grip.

CS-6 The Sound Grip comes with a stand for enhancing your PSP movie-watching experience.

CS-7 There are a separate power switch and volume control on the Sound Grip.

CS-8 A PSP safely locked into place. There is DC power input and headphone jack on the back of the Sound Grip.

CS-9 A clear cover holds the PSP in place and a stand holds the Sound Grip at an optimal angle for movie watching—hands free.

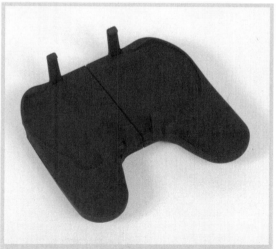

CS-10 Boxwave Corporation PSP ActionGrip.

CS-11 Yup, when you attach the ActionGrip to your PSP, it sure does look a lot like the PS2 game controller. And, familiarity breeds success, at least for gamers looking for that PSP gaming edge.

HOW TO MOD YOUR LOGITECH PLAYGEAR POCKET WITH A CUSTOM SKIN

PPS-1 Logitech PlayGear Pocket.

PPS-2 Open up the PlayGear Pocket and set your PSP inside.

PPS-3 The top lid of the PlayGear Pocket hinges up for access to your PSP and also shields the screen from the sun.

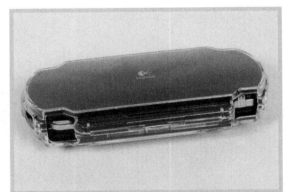

PPS-4 When the PlayGear Pocket is closed, you can still charge your PSP and listen to music. Both of those attributes are extremely valuable to the PSP that's on the go.

PPS-5 You can add your own custom skin to the inside cover of the PlayGear Pocket case.

PPS-6 Gently pull out the rubber liner from the inside of the case, insert your skin, and replace the liner.

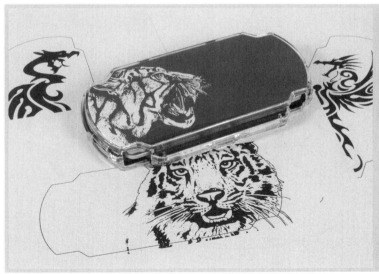

PPS-7 A wide variety of skins can be quickly and easily interchanged with the PlayGear Pocket case.

PPS-8 **Use this blank template for designing your own PlayGear Pocket skin.**

PPS-9 Here is a ready-made skin for your PSP. Cut this template out of the book and insert it into your PlayGear Pocket. I would not recommend making a photocopy of this template or using a laser printer to make your own templates. Why? The toner on the photocopy and laser print can stick or fuse onto the case's plastic and can become very difficult to remove. Therefore, use either the template in this book or make your own templates with an ink-jet printer.

CHAPTER 9

Get a Wi-Fi of This

Building WLAN into the PSP is a quantum leap over the PS2 (Fig. 9-1). Even the newer (and much improved) SCPH-7001x version of the PS2 lacks a workable hardware solution for adding Wi-Fi networking to the game console.

Oh, sure, including the Ethernet/modem network capability in the newer PS2 was a long-overdue addition, but game consoles are traditionally housed near TV sets and not near computer Ethernet hubs. As PS2 gets long in the tooth, it's time to breathe one last breath of life into this aging classic gaming console.

This is a simple hack that will bring your PS2 up to speed with your PSP. By converting the PS2 Ethernet port into an active Wi-Fi node, you can access the same WLAN that you use with your PSP.

FIGURE 9-1
Wireless networking is built into the PSP, making point-to-point and Internet access a snap.

145

CHEAT CODE: Turn your PSP into a Wi-Fi sniffer. Just follow these simple steps to create a simple wardriving system.

1. Highlight "Network Settings" from the main "Settings" menu and press the X button.

2. Highlight "Infrastructure Mode" and press the X button.

3. Determine a name for this new network connection, such as "PSPWardriver," then press the X button. The built-in PSP keyboard will be displayed for typing in your selected name.

4. Press the right arrow button to accept the network connection name and move onto the next screen.

5. Press the up arrow key to move up to the "Scan" option and press the X button (if needed). Your wardriving Wi-Fi snooping PSP will now search for all access points that are within its wireless network radius.

6. A list of wireless access points along with their SSIDs and WEP key status will be displayed on your PSP (Fig. 9-2). What you do with this newfound power is your responsibility.

FIGURE 9-2 Believe it or not, this is an actual wardriving PSP scan. It took me about 2 minutes to find this open access area. Sweet. It's a big wide open Wi-Fi world out there. Even better, a group of wardrivers, who shall remain nameless, scanning a city for Wi-Fi open networks marked their results on street curbs in chalk. This action was so strange to the local residents in this Midwestern city, that the local police and regional FBI agents began searching for the terrorists who were making these markings. Weird.

Sever the Tether

How the heck can you add Wi-Fi network connections to a device that only has an Ethernet port? Simple, find a wireless-Ethernet bridge. One such bridge is the Linksys Wireless-B Game Adapter (model number WGA11B). Best of all, this thing is a snap to set up.

Find a cozy spot for your Game Adapter: typically, right next to your PS2. Plug the power supply in and use the supplied Ethernet cable to connect the Game Adapter to the PS2. If you have a problem talking between the Game Adapter and the PS2, there is a crossover switch located on the back of the Game Adapter (it is labeled X-II). Toggle this switch until a handshake is obtained.

If you are going to play another PS2 in an ad hoc network competition, then simply match the channel selector between the two consoles. However, if you wish to game online, then set the Game Adapter to the Internet channel (e.g., In). That's it—you be fragging.

There is one precaution that could prevent your jacking into the Internet. The Game Adapter is set up for wireless networks that have no WEP encryption settings and a readable SSID. This is where your PSP can come in handy. Use the PSP in its Wi-Fi sniffing or wardriving configuration (as described elsewhere in this chapter) to locate and identify wireless networks that match these requirements. If your network is protected from outside use through WEP encryption, then you will have to establish an Ethernet connection between your PC and the Game Adapter for manually configuring your Wi-Fi game play. A setup wizard program is supplied with the Game Adapter for simplifying this stupefying process.

> **CHEAT CODE:** Are you always looking for a Wi-Fi hotspot and can't seem to find one? Then you need a Wi-Fi snooper or sniffer. Devices such as the WiFi Seeker enable you to quickly determine the presence of a Wi-Fi hotspot, usually with just the press of a button.

Oh, and if you're looking for a good wireless router, then look no further than the Apple Airport Extreme Base Station (Fig. 9-3). If you don't believe me, then consider this real-life scenario about the Airport's reliability.

As a tribute to the robust nature of the Airport Extreme Base Station, let me tell you how my system survived Hurricane Dennis. Although I didn't live

HOW TO SET UP A WI-FI NETWORK CONNECTION FOR YOUR PS2

LS-1 You can add the Wi-Fi connectivity to your PS2 with the Linksys Wireless-B Game Adapter.

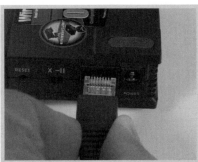

LS-2 Begin your Wi-Fi installation by connecting the supplied Ethernet cable to the Linksys Game Adapter.

LS-3 Plug the other end of the supplied Ethernet cable into your PS2.

LS-4 Select either a wireless channel for ad hoc connectivity or establish an Internet connection just by pressing the channel selector button.

LS-5 If your wireless network has WEP key security encryption, then you will have to use the supplied setup wizard for configuring your Linksys Game Adapter. This setup program must be run from a network connected PC.

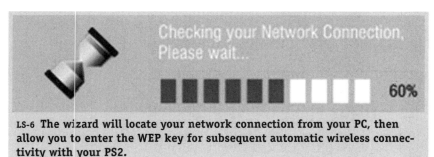

LS-6 The wizard will locate your network connection from your PC, then allow you to enter the WEP key for subsequent automatic wireless connectivity with your PS2.

in the exact path of this hurricane, I did suffer one of the exciting attributes of such storms—lightning. On July 10, 2005, as Hurricane Dennis waded ashore near Pensacola, Florida, a band of significant rain began swamping my area. During this deluge, a bolt of lightning struck a tree in my backyard. As luck would have it, this tree was adjacent to my buried telephone line. Somehow, the lightning sent a cloud of steam up into the air, blew the bark off of the tree, and sent a pulse of high voltage along the telephone line and inside my house—all in the blink of an eye.

After the storm, I tried to access my network, and the entire system was dead. I examined my network room and discovered that the Airport Extreme Base Station was dead (i.e., the power indicator was off), the Ethernet was down, and the network printer was offline. Remarkably, in about 5 minutes, I was back online. The Airport Extreme Base Station just needed to be unplugged for 5 minutes and plugged back into its power supply. The network printer was then restarted by the Airport system. Unfortunately, the Ethernet network was inoperable. New cable had to be pulled, and then all was right in my little universe, again. Thanks to the resilience of the Airport Extreme Base Station.

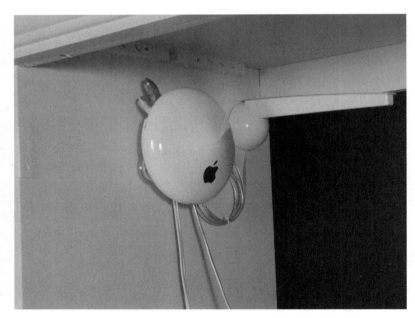

FIGURE 9-3 **You can sequester an Apple Computer Airport Extreme Base Station just about anywhere. Here, the Base Station has been mounted inside an armoire along with a Dr. Bott ExtendAir Omni Antenna. The addition of this external antenna extends the range of the Airport wireless signal to more than 150 feet with a 360-degree coverage—which is more than enough range for allowing me to write this caption outside next to the pool.**

EASTER EGG: OK, PSP is portable, but Wi-Fi is not. If you're not around a Wi-Fi hotspot, you can't jack in. Too bad there isn't some sort of junction box or something that would enable you to turn your cell phone into a portable wireless transmitter. Well, guess what? There is the Junxion Box (Fig. 9-4).

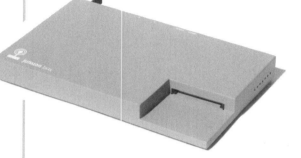

FIGURE 9-4 The Junxion Box JB-110b. A PC Card modem must be added to this configuration before you can establish a wireless WAN. (Photograph courtesy of Junxion, Inc.)

The Junxion Box is a wireless wide-area network (WAN) router that takes a standard PCMCIA (or PC Card) modem and connects it to any Wi-Fi-certified appliance. Regardless of operating system (Linux, Mac, Windows) and regardless of hardware (desktop, notebook, or PSP), the Junxion Box is able to get you connected anywhere; anywhere that your PC Card modem can catch a satellite signal, that is.

It is virtually magic, in its ability to completely insulate you from all of the typical connectivity issues that can turn the computer age into the stone age. Just pop a PC Card modem into the Junxion Box, activate and configure the network, then access a cellular network with any device via Wi-Fi connectivity.

Oh, and if your hardware device lacks Wi-Fi capability (e.g., a stock PS2), the Junxion Box can accept an Ethernet connection, too. Likewise, once you're jacked into this cellular network via the Junxion Box, you can access the Internet, share files, and print to a network printer simply by connecting to the Junxion Box and setting up its configuration via a Web browser.

You can select your PC card modem and its associated cellular service from several major vendors, including: Alltel, Cingular, Sprint, and Verizon Wireless. Once activated and connected, your Junxion Box network router can enable your PSP to easily access Web portals via the *Wipeout Pure* browser hack (see Chapter 2).

HOW TO SET UP A WI-FI NETWORK ON YOUR PSP

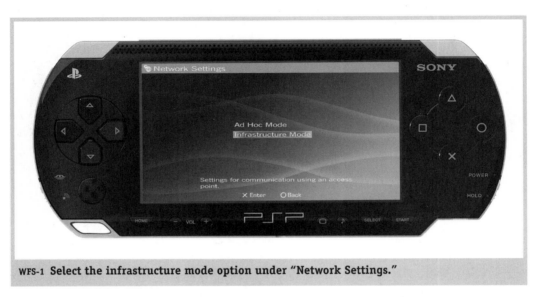

WFS-1 Select the infrastructure mode option under "Network Settings."

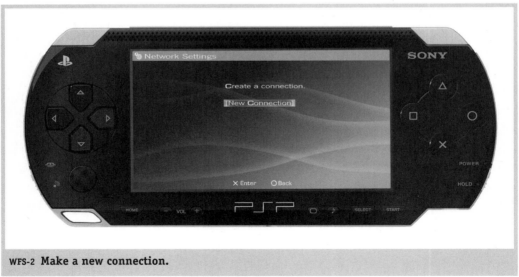

WFS-2 Make a new connection.

WFS-3 You can assign any name that you wish to this new connection. Likewise, you can make several different connections based on various network configurations. So, use very clear and descriptive names.

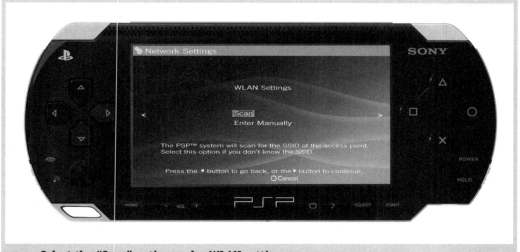

WFS-4 Select the "Scan" option under WLAN settings.

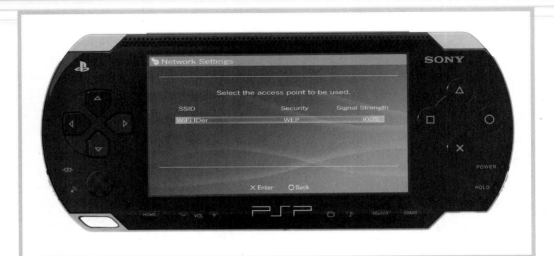

WFS-5 Your PSP will return a listing of all Wi-Fi access points that are within wireless range of your system. Select your preferred access point.

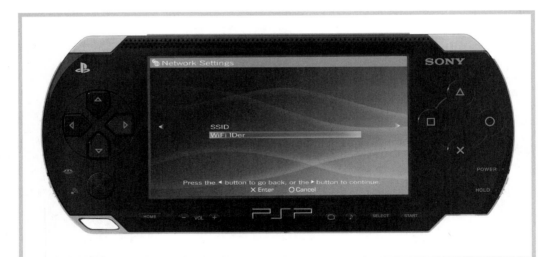

WFS-6 Verify that this is the SSID for your preferred access point and accept this entry by pressing the right arrow button.

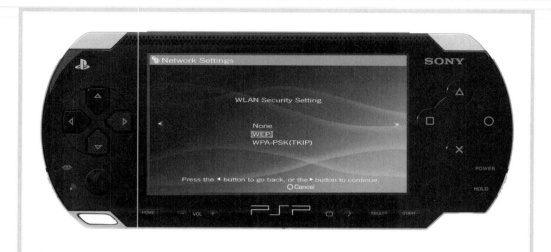

WFS-7 Select the correct security option for your preferred access point under the WLAN security setting.

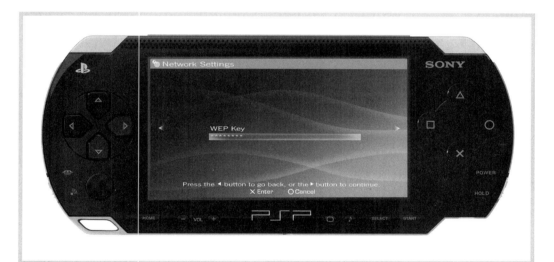

WFS-8 If required by your preferred access point, key in the WEP encryption key. Although this could be a 26-character key, the PSP will only display 8 asterisks.

WFS-9 Select "Easy" under "Address Settings." Accept the settings displayed by the system and test your WLAN connection. Now go do some surfin', dude.

CHAPTER 10
USB OTG— Let's Go

Sony made all of the right moves with the PSP. With USB, Wi-Fi, Memory Stick PRO Duo, and UMD in its design, your PSP is well choreographed for waltzing around with some cutting-edge media technology. Unfortunately, most of the time it's like you're dancing in the dark—you really don't know much about any of this media.

Sure, you can buy UMD movies and games. Likewise, I've already shown you how to access your photo, music, and video collections (see Chapter 1) and play homebrew games (see Chapter 7) all from Memory Stick PRO Duo media cards that have been loaded via USB. Other than that rather short laundry list of media machinations, the PSP media remains tightly controlled and cloaked behind specifically defined, restricted-access portals.

All is not lost, however. As the popularity of the PSP grows, the depth and breadth of accessories and peripherals that can more fully exercise each of these various media will increase. Until that time, we hackers must exercise our own Yankee ingenuity. We must attempt to integrate the PSP into our own digital lifestyles.

One of the most promising avenues for exploiting PSP media that I've stumbled across is the USB port (Fig. 10-1). Let's face it, this darn interface is everywhere and on everything. And seeing it on the PSP makes you feel that just

FIGURE 10-1
USB pin outs.

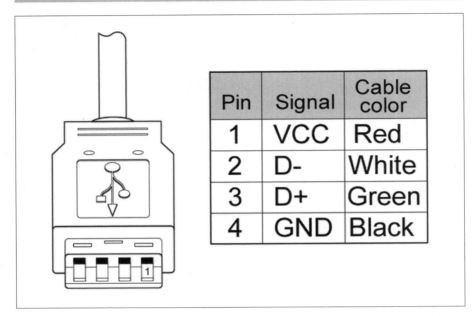

about anything must be possible to connect to it: keyboards, external hard disk drives, cameras, printers, DVD players, and CD-R burners. Alas, this wishful thinking is nothing more than a pipe dream.

After only a cursory examination, you'll realize that the PSP-USB port is buttoned up...tight. For example, the port's activation is controlled via firmware inside the "Settings" menu with the "USB Connection" option. There is a faint glimmer of hope that USB on-the-go (OTG) might be the answer to using the PSP-USB port to *your* advantage. What is USB OTG? you ask.

Let's Go Mobile

The USB Implementers Forum gave road warriors an early Christmas present in 2001, with the announcement of USB OTG. As an extension of the USB 2.0 specification, USB OTG enabled specially configured USB peripherals to become hosts for connecting to other USB peripherals. Traditionally, this host context was provided by a PC. By removing the computer from this peripheral connectivity equation, USB OTG endowed peripherals with the ability to act as hosts. Now your mobile stuff is truly mobile.

By using a USB OTG peripheral, you can plug in a digital camera, for example, and download images onto an external storage device. Or, you can use a USB OTG peripheral for connecting a handheld device to a cell phone for accessing and purchasing music via a cellular connection. Thankfully, a USB OTG host does *not* require that the other connected device either support or recognize USB OTG. All that is required is that the peripheral must be a USB device. That's it.

Will USB OTG work with your PSP? Well, that's the $64,000 question. Beware that not all USB OTG hosts are created equal. For example, when it was first released, the Macally SYNCBOX II would *not* work with the PSP. Therefore, make sure you thoroughly evaluate a USB OTG peripheral for PSP compatibility before you purchase it.

If you read just a little further in this chapter, I will tell you about the world's best USB OTG product for working with your PSP. It's the best hack ever thunk up by me. You'll love it. Just keep reading. OK, if you can't wait, just skip down to the section USB OTG PSP.

Get Pumped

Another aspect of the USB port that is finally being more fully used by commercial products is the ability to derive +5 volts from pin 1 of the USB connector. This ability enables the USB port to power connected peripherals. Or, in the case of this neat little hack, charge your PSP.

Basically, your computer's USB port will provide +5 volts to a connected PSP. If a barrel plug adapter is substituted for the mini B connector on a standard PSP USB cable, then you can have your computer charge your PSP. Simple enough, right? Wait, there's more. Rather than just throw away that valuable mini B connector, you can take the two data lines from this same USB port on your computer (e.g., pins 2 and 3) and send data from your computer to your PSP while you are charging it. All of this magic, at the same time, from the same port, on the same cable.

> **CHEAT CODE:** If you do a lot of USB hacking, then sooner or later you will need the free USB-View utility. This utility comes in two versions: one for Windows and one for Linux. You can download both versions from the FTDI Web site (www.ftdichip.com).

HOW TO BUILD YOUR OWN USB PSP CHARGER

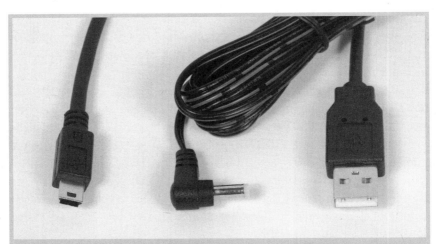

UC-1 Take three cables, some heat-shrink tubing, and a modest amount of soldering, and you can charge your PSP while you are accessing your Memory Stick PRO Duo media cards; at the same time. I used Digi-Key part numbers AE1292-ND and CP-2198-ND for my cables. Make sure that you run the power lines from the USB-A plug to the charger plug and the data lines to the USB-B plug.

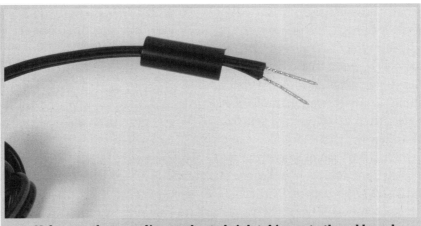

UC-2 Make sure that you slip your heat-shrink tubing onto the cable ends prior to soldering.

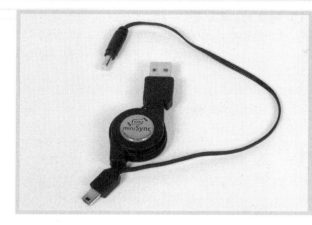

UC-3 Boxwave Corporation makes a miniSync cable that will perform the same charging/USB access task.

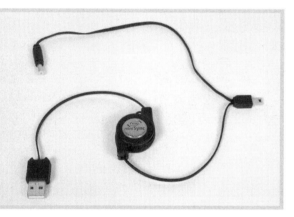

UC-4 Pull each plug of the miniSync cable out until they stop.

UC-5 Hook 'em up. Charging rates for both the DIY version and the Boxwave product take about the same amount of time as the Sony-supplied charger— 2 to 3 hours.

First, you must accept this warning and caveat regarding this hack. My measurements of the supplied Sony battery charger indicates that both +5 volts and 2 amps are juiced into the PSP, whereas the USB charger is only able to produce +5 volts and 500 milliamps. Therefore, the PSP battery will not be charged at the optimal current specified by Sony. Yes, this will *void* your warranty, but, no, it will *not* hurt your PSP battery. I have used this arrangement for recharging several PSP batteries. On comparing the USB-charged batteries to the Sony-charged batteries, I have not detected any significant differences in battery capacity or life span. So with that little bit of hand-wringing out of the way, see the previous two pages for some cable cutting and splicing.

USB OTG PSP

If you're looking for the Holy Grail of USB OTG functionality with your PSP, then look no further than the Cowon America iAUDIO X5. Sure, this book is packed with cool hacks, mods, and expansions for the PSP, but the iAUDIO X5 is one of the best hacks that you can do with your PSP—this is a killer hack.

By using one simple cable (that's provided with the iAUDIO X5), you can leave your notebook computer at home and go on the road with *just* your PSP and an iAUDIO X5. That's it, nothing else. The iAUDIO X5 is capable of being your bridge between *all* of your other electronic lifestyle products and your PSP. It is the essential missing element that turns the PSP into the most powerful media player in the world.

Like a modern Rosetta stone (no, not that Rosetta, used for emulating Intel-based Mac OS X on PowerPC processors), the iAUDIO X5 is capable of linking your PSP to hundreds of different USB products for transferring files, viewing images and text, listening to music, and watching videos. Configured with a 20-GB hard disk drive, the iAUDIO X5 is the perfect PSP companion.

Because I consider the iAUDIO X5 to be an indispensable device for endowing the PSP with capabilities that Sony hasn't even dreamed of, let me, first of all, tell you where you can buy your own iAUDIO X5. I have found the iAUDIO X5 at Target, as well as at Earth's largest store—Amazon.com. In both cases, the manufacturer's name was JetAudio, Inc. rather than Cowon America. No matter; the value of the iAUDIO X5 for PSP owners is beyond reproach.

iAUDIO X5 Features

If you still don't think that the iAUDIO X5 is the killer hack for the PSP, then consider these unbelievable *other* features:

- MP3
- OGG
- Windows Media Audio (WMA)
- ASF
- FLAC
- WAV
- MPEG4 (video) playback (must convert using jetAudio)
- FM radio receiver and recording
- Voice recording
- Line-in recording
- TXT (text)
- JPEG file viewer (image enlargement, background screen designation)
- Built-in HDD (20 GB or, optionally, 30 GB), USB host, file copy/delete
- USB-2.0 interface with OTG-host capability
- Resume, fade-in, autoplay features
- Search speed, skip speed setup
- Volume (digital 40 levels)
- Various EQ and sound field effects
- Clock, alarm, timer recording, sleep function, power-saving shut down
- LCD auto-off time adjustment, brightness and contrast adjustments
- Battery—built-in rechargeable lithium-ion battery (up to 14 hours continuous playback); charging time approximately 3 hours
- Buttons—five-way jog joystick (NAVI/MENU, VOL+, VOL-, REW, FF), PLAY, REC, POWER & HOLD switch
- Display—160- × 128-dot, 260,000-color TFT LCD
- Size (excluding the LCD)—4.08 × 2.39 × 0.56 inches (103.7 × 60.8 × 14.3 mm)
- Weight (including the built-in battery)—5.11 oz (145 g)
- Firmware upgrade
- Supports Mac OS X (data transfer only)
- Supports Linux v. 2.2 or higher (data transfer only)

HOW TO USE USB OTG WITH YOUR PSP

U0-1 The Holy Grail of PSP connectivity—Cowon America iAUDIO X5 plus the supplied USB-host cable.

U0-2 The USB-host port is different from the standard iAUDIO X5 USB port.

U0-3 Hook everything together, power up the iAUDIO X5 and the PSP, and activate the PSP USB connection.

U0-4 Once you're in the iAUDIO X5 host mode, press the REC button to switch the connection to the PSP.

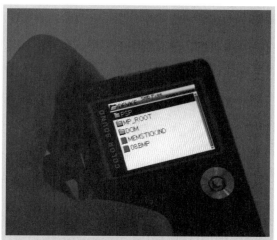
U0-5 Voilà, you're on the PSP Memory Stick PRO Duo media card without a computer. And the angels did sing.

U0-6 Navigate through your PSP folders with the iAUDIO X5 multifunction joystick jog button.

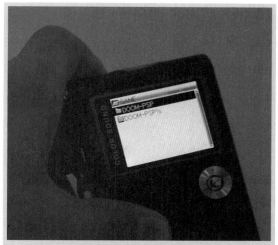
U0-7 You can copy, move, delete, and manage all of your homebrew games with the iAUDIO X5.

U0-8 You can also copy, move, delete, and manage all of your images with the iAUDIO X5.

U0-9 Yes, you can see Mac OS invisible files on the iAUDIO X5.

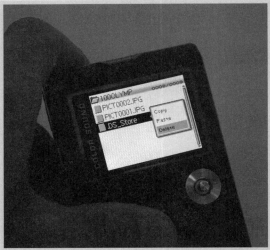

U0-10 And, you can delete Mac OS invisible files with two simple clicks on the iAUDIO X5 joystick.

U0-11 Locate an image file that you want to manage.

U0-12 Two clicks on the iAUDIO X5 joystick and you can copy an image into the host's memory.

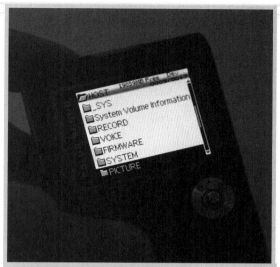

U0-13 **Navigate to the PICTURE folder on the iAUDIO X5.**

U0-14 **Paste the copied image onto the host's 20-GB hard disk drive. The unique "PasteIn" command actually pastes the selected file inside the target folder without opening the folder. Now that is crackin'.**

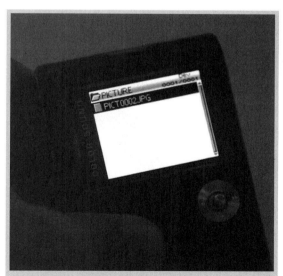

U0-15 **You have just copied an image from your PSP to the iAUDIO X5. Wow!**

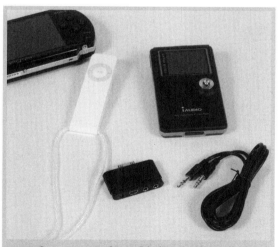

U0-16 **Can you see where this is going? You bet, iAUDIO X5 is the perfect missing link for connecting your PSP to your iPod shuffle. You could mess with the line-in port for hooking the two up.**

U0-17 A smarter alternative, though, is to use the USB host capability of the iAUDIO X5.

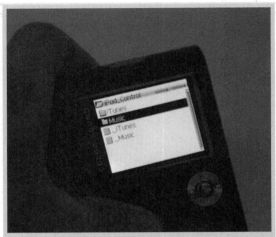

U0-18 There we are: inside the iPod shuffle. Who needs iTunes to (mis)manage our music?

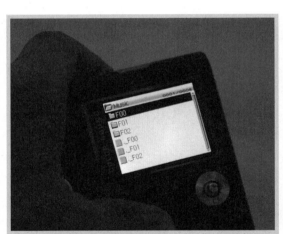

U0-19 Navigate through the iPod shuffle folder hierarchy with the iAUDIO X5 joystick.

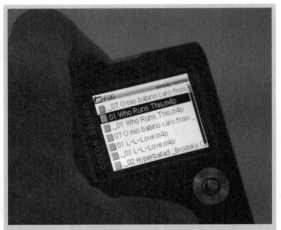

U0-20 There's your music. Note: Yes, these are DRM-locked M4P tracks and, therefore, they won't work on your PSP. Always use MP3 music and you'll never have a problem shuttling your shuffle music to your PSP. Also, notice how the Mac OS created invisible files for every music track. Naughty Mac OS.

U0-21 Other devices can also benefit from the iAUDIO X5 host mode. This Aiptek IS-DV video camcorder is the perfect video companion for your PSP.

U0-22 After shooting some video tracks, just upload them to the iAUDIO X5. Unfortunately, these tracks are compressed in the ugly ASF file format and can't be run directly on the PSP. Rats. By using the host function of the iAUDIO X5, however, you can keep on filming until you're back at your computer for converting the ASF tracks into PSP-friendly MPEG4 movies. In this capacity, you could shoot up to 20 GB worth of footage with the IS-DV camcorder. Wow!

U0-23 Looking for taking some still photographs with some style? The Philips GoGear Wearable Digital Camcorder is great for capturing those candid moments in style.

U0-24 In just a couple of moments, you can have your JPEGs from the GoGear camcorder onto your PSP for showing to all of your friends. No muss, no fuss; just click, upload, download, and enjoy.

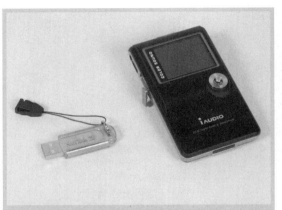

U0-25 If you've got a USB storage device, you can hook it up to the iAUDIO X5 to give you USB connectivity freedom.

U0-26 Yes, unfortunately, you can't do much with these ASF tracks with a remote USB host. Rather, use the host as a temporary holding device, so that you can keep on filming. Then convert the ASF tracks into something more usable when you return to your PC.

U0-27 Keep all of your music in MP3 format and you can use the iAUDIO X5 as a jukebox for selecting, moving, and managing your entire music library with all of your USB-equipped consumer electronics music products.

U0-28 Three screens living in perfect harmony. Even products such as the Audiovox PVR1000 can be integrated into your PSP universe with the iAUDIO X5.

- → Included software (Windows only): JetShell (file transfer, MP3/WMA/WAV/audio CD play, MP3 encoding); JetAudio (integrated multimedia player software, music/video conversion)
- → Black-brushed anodized aluminum exterior
- → Fully compatible with PSP

I have hacked several potent USB products into the iAUDIO X5 for jacking into a PSP. See the previous seven pages for how.

A Stupid USB Trick

If you own *both* the PSP *Metal Gear Acid,* Japanese version, and the PS2 *Metal Gear Solid 3,* Japanese version, you can do a little USB trickery. Just plug your USB cable into both the PSP and the PS2 and fire up both games. Inside the PS2 version, you will be able to get, save, and use the Stealth Camo and EZ Gun from the PSP version. Make sure that you save the PS2 game prior to disconnecting the PSP. Then you will have both of these weapons for use in future PS2 game play with *Metal Gear Acid Solid 3* (Japanese).

CHAPTER 11

Status Quo Vadis?

Sony isn't the first manufacturer to offer you a digital lifestyle. Products from Aiptek, Apple Computer, Audiovox, Cowon Systems, Olympus, Mattel, and Tiger Electronics have each claimed to have the ideal solution to your living in a digital world.

Hack Not What You'd Want in a Camera

If you follow the world of hackers very much, then you probably heard all of the buzz about the "disposable" video camera manufactured by Pure Digital Technologies and distributed by CVS Pharmacy that had been hacked into being a "reusable" video camera. Well, so what? This $30 camera isn't worth the effort that was expended to make it into a reusable video camera. Its specs are appalling, and its performance is disappointing. A much more acceptable solution to low-cost video is the Aiptek IS-DV (Fig. 11-1).

This palm-sized, cute camera packs an impressive punch for less than $160. Rather than using video tape for recording video, the IS-DV uses SD/MMC media cards (Fig. 11-2). Therefore, you can shoot until you fill a card, swap it out, and keep right on filming. Try that with the CVS video camera.

FIGURE 11-1
The Aiptek IS-DV comes with everything you could ask for in a low-cost camcorder, even a tripod.

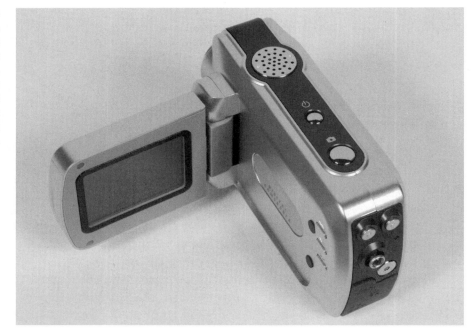

FIGURE 11-2
The Aiptek IS-DV does not use videotape. All video tracks are recorded on SD/MMC media cards.

In fact, the complete specs for the Aiptek IS-DV are equally impressive. Consider this:

- Camera resolution—2560 × 1920 pixels (5 megapixels)
- Video resolution—2048 × 1536 pixels (3 megapixels); 640 × 480 pixels (VGA) up to 30 fps; 320 × 240 pixels (QVGA) up to 30 fps
- LCD—1.8-inch LTPS color-LCD display
- Digital zoom—4×
- Internal memory—16-MB Flash memory
- AV output
- Image format—JPEG, ASF
- Focus range—two settings—macro, 40 cm; normal, 250 cm to infinity
- Exposure control—auto; EV, +2 ~ −2
- White balance—five modes
- Still image effect—four modes
- Interface—USB 2.0
- Power source—NP60 lithium-ion rechargeable battery, or via USB port when plugged into PC
- Camera dimensions—3.86 × 2.57 × 1.28 inches
- External memory—SD/MMC card slot (maximum 1 GB)
- Built-in microphone
- Built-in speaker

One of the biggest sticklers with the IS-DV, however, is the recording of ASF tracks on the camcorder. These tracks are virtually worthless on every professional video playback system, including the PSP. And don't tell me about the Microsoft Media Player; I said *professional* video playback, not pathetic video playback. In the case of the PSP, these ASF tracks must be converted into MPEG4 movies. Luckily for Aiptek owners, the IS-DV comes packaged with a nifty little program for converting ASF into AVI. Then, these AVI files can be easily made into MPEG4 movies. Unfortunately, for Mac OS X PSP owners, there is no easy way to convert ASF into MPEG4.

HOW TO CONVERT ASF TRACKS INTO MPEG4 MOVIES

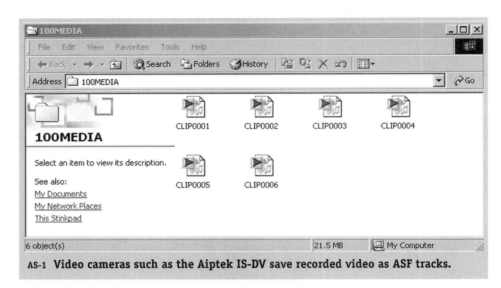

AS-1 Video cameras such as the Aiptek IS-DV save recorded video as ASF tracks.

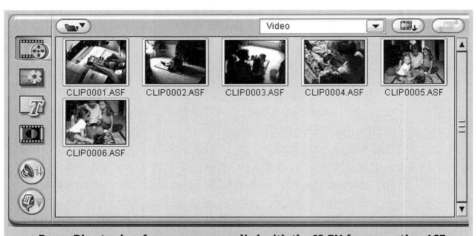

AS-2 Power Director is a free program supplied with the IS-DV for converting ASF tracks into AVI files.

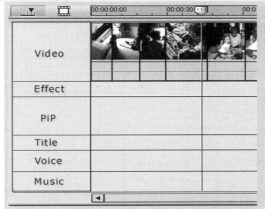

AS-3 Drop your ASF tracks into the Power Director timeline.

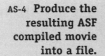

AS-4 Produce the resulting ASF compiled movie into a file.

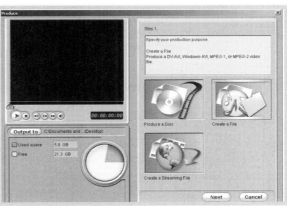

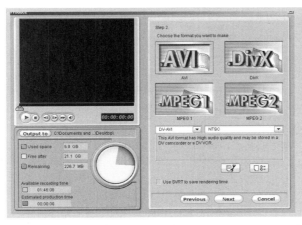

AS-5 Only AVI format files can be created with the *free* version of Power Director. If you want more format options, then upgrade your copy of Power Director.

Apple Bites Sony Where It Hurts

As unbelievable as it sounds, a computer manufacturer designed a product that can help alleviate some of the boredom associated with life. The Apple iPod mini (Fig. 11-3) is a svelte 3.6 by 2.0 by 0.5 inches, 3.6-ounce anodized-aluminum digital music player with a 4-GB or 6-GB hard disk drive that Apple claims can hold 1000 and 1500 songs, respectively. Additionally, the iPod mini comes with a dual function FireWire (and USB 2.0) cable that can both charge the Pod's battery, as well as shuttle music back and forth from the included iTunes® Music Store software. The iTunes software acts as a digital music jukebox that is capable of buying digital music from the Apple Music Store, converting CD discs into iPod-aware advanced audio coding (AAC) encoded music, and managing your digital music database (e.g., copying music to the iPod mini, creating playlists, deleting music from the iPod mini, etc.).

FIGURE 11-3 **Apple iPod mini. (Photograph courtesy of Apple)**

A patent-pending, touch-sensitive, solid-state Click Wheel is used for scrolling through all of the iPod mini's functions and for selecting and playing songs from your loaded music collection. Rounding out the iPod mini is a reported 8-hour battery life, 1.67-inch monochrome LCD with LED backlight, earbud headphones, AC adapter, and belt clip. Finally, you can make your own fashion statement by selecting the best Pod for your hacking persona—silver, gold, pink, blue, or green colors are available. In initial shipments of the iPod mini, the blue anodized aluminum model was the front leader in advance sales.

Even though Apple based its storage capacity claim on a 4-minute-per-song file size, my testing of the iPod mini 4-GB version, for example, resulted in a random, real-world music sampling being able to store 946 songs on the 4-GB hard disk drive. In fact, if your music library contains a large number of longer length recordings (e.g., classical music), you should expect your iPod mini to hold even fewer songs.

Likewise, my playback tests could only squeeze 4 1/2 to 5 hours' worth of music out of a fully charged iPod mini. Although the battery strength indicator did display a reduced charge, no other type of warning message was displayed prior to the iPod mini shutting down from lack of power. (Note: In fact, whereas the battery strength icon was showing a completely exhausted battery, I was still able to listen to Green Day's "American Idiot" in its entirety before my 'Pod went dead—that's more than 45 minutes' worth of music. A subsequent attempt at starting up the iPod mini, however, did display a brief message stating that the 'Pod was out of power.) Recharging the completely drained iPod mini battery took 2 hours and 15 minutes.

Close Enough for Hand Grenades and Horseshoes

Remember that kid in school who always tried real hard but could never win the big race? Well, in consumer electronics lifestyle products, that kid is Audiovox Corporation. Sure, the specs for its products look great on paper, but Audiovox products never seem to be well known among electronics aficionados. Take the Personal Video Recorder PVR1000 (Fig. 11-4), for example.

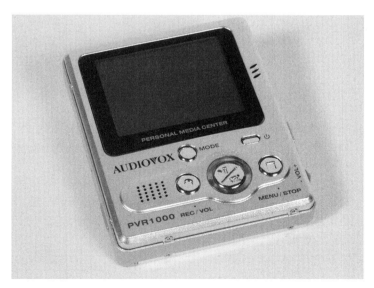

FIGURE 11-4 Audiovox PVR1000.

Yes, you can use the PVR1000 with your PSP, but there are many other products that deliver the same goods. The glowing virtue of the PVR1000, though, is that it has a big LCD screen *and* a set of audio/video (A/V) inputs. Now, don't mistake the presence of these A/V inputs as any implication that you will be able to hook it up to your PSP for external movie/game viewing. Because you can't. Rather, you can repurpose the PVR1000 for hooking it up to a game console such as a PS2 for remote game play (see Appendix A).

PVR1000 Features

If only; you'll find yourself saying that a lot with the PVR1000: If only it were better built (e.g., I sampled three products, and they suffered from a bad battery compartment door, a stuck on/off button, and a misaligned A/V input port). If only it had a better resolution screen (e.g., the LCD can get dirty *underneath* the protective outer screen). If only it would have better performance (e.g., the sound quality is mediocre through the built-in speaker). Nonetheless, if you can find one cheap (e.g., less than $80), grab it. The parts and features are great for repurposing with other hacks and projects. Oh, and if you ever see the message "PVR ERROR" appear on your LCD screen, this product is totally dead. A doorstop. I had one PVR1000 display this informative screen right out of the box. Lovely. Get this, though. The kind folks at Audiovox actually replaced the dead PVR1000 at no charge in less than one week. Now that's customer service.

- 32-MB on-board memory
- SD/MMC memory; 1 GB maximum
- USB interface
- Digital photo player, digital video player, digital video recorder, digital voice player, digital voice recorder, MP3 player
- Playable media files—ASF (MPEG4 video), JPEG (photo), MP3/WAV (audio) (Note: All files must be converted with included software.)
- 2.5-inch color TFT LCD
- One speaker
- Microphone—built-in condenser
- Quality settings—voice recording, audio/video recording—low, high

- ➜ TV system—NTSC/PAL
- ➜ Clock and calendar
- ➜ Games included—tic-tac-toe, puzzle
- ➜ Direct printer services (DPS) supported
- ➜ A/V input—1 minijack: 1/8-inch composite-video + audio
- ➜ Headphone jack—yes (0.125 inch)
- ➜ Hold switch—yes
- ➜ Media-playing compatibility—Windows 98/2000/ME/XP
- ➜ Mass-storage compatibility—Windows 98SE/2000/ME/XP; Macintosh OS 9.X and 10.1 and above
- ➜ Power source—battery or AC/DC power adapter (supplied)
- ➜ Batteries—rechargeable lithium-ion ×1 (internal/supplied)
- ➜ Battery life—information not available
- ➜ Dimensions—3 × 3.7 × 0.78 inches
- ➜ Weight—0.25 pound (without battery)

Would You Like a Little Video with That iPod?

What if you were able to beat Steve Jobs to the punch and integrate video alongside the now famous iPod sound? Well, Cowon Systems has already done that with the iAUDIO X5.

The iAUDIO X5 sports a 1.8-inch, color LCD that is capable of displaying digital photographs, music videos, and movies. Oh, and did I say that it can play music, too? Available with either 20 GB or 30 GB of memory capacity (note: this is a hard disk drive system), the iAUDIO X5 can outperform the entire iPod line, leaving Apple's music players behind in a cloud of dust.

> **CHEAT CODE:** See Chapter 10 for more information about using the iAUDIO X5.

Although holding the aluminum iAUDIO X5 is a sheer joy, this beauty is more than skin deep. After just a short 3-hour charge, the iAUDIO X5 is able to belt out music for more than 14 hours of use. Now, throw in a multifunction navigation joystick, a USB OTG host capability, a high-speed USB-2.0 port, and a direct audio MP3 encoding ability (that's the ability to rip a CD without using a computer), and iAUDIO X5 becomes the perfect PSP companion.

HOW TO USE THE iAUDIO X5 AS THE PERFECT PSP COMPANION

CS-1 The Cowon America iAUDIO X5 can be easily adapted for use with your PSP.

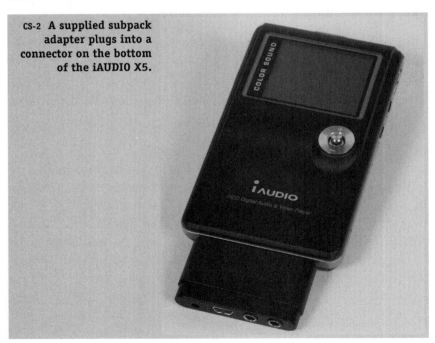

CS-2 A supplied subpack adapter plugs into a connector on the bottom of the iAUDIO X5.

CS-3 The subpack adapter provides the iAUDIO X5 with a charger connector, a USB port, and a line-in jack. In fact, the line-in jack will allow you to rip *any* external sound source into an MP3 format file.

CS-4 Hidden away on the side of the iAUDIO X5 is its best feature—a USB-host port for real live USB OTG capability.

CS-5 A supplied USB-host cable will connect the iAUDIO X5 to your PSP for über file management. You can now master your PSP anywhere, anytime without a computer. See Chapter 10 for more information about this outstanding capability.

Cowon Systems also includes a digital multimedia mastering application called JetAudio that can easily translate all of your media files (e.g., pix, tunes, and flicks) into supported iAUDIO X5 file formats (e.g., MP3, WMA, ASF, WAV, OGG, and the FLAC lossless audio compression codec, as well as JPEG for stills, and XviD and MPEG4 for video).

Even if it does have a goofy name, the iAUDIO X5 will endear itself to your heart (and ears and eyes) with an alarm system, time-shift recording, voice recorder, and mobile storage facilities. Couple this impressive feature list with Linux, Mac, and Windows support and the iAUDIO X5 is the perfect mate for your PSP.

Also, be sure to keep an eye on Cowon Systems. The upcoming release of its next-generation Cowon A2 portable multimedia player with a 4-inch, 16:9 wide-screen LCD is a state-of-the-art product that could set the standard for all future handheld entertainment systems. Couple the A2 with an ability to record TV programming directly onboard this player and you will have made a bold step toward a truly digital lifestyle. That is, unless the entertainment industry doesn't cripple it first.

What the Heck Was That Name Again?

What is it with some of these new digital product names? Case in point: the Olympus m:robe MR-500i. Boy, that just rolls off your tongue, doesn't it? Hidden behind one of today's weirdest product names is a 20-GB hard drive, 1.22-megapixel camera, 3.7-inch, color VGA LCD touch-screen music player thingie.

Although there are a couple of worthwhile design points with the MR-500i that bear further mention (e.g., onboard camera and synching a photo slideshow to your music), the touch screen isn't one of them. In just a few short minutes of use, the screen becomes completely smudged from your fingers. Even worse, this same touch-screen interface is used for activating and firing the digital camera's shutter release. Ironically, in advertising the MR-500i, Olympus attempts to put a good face on the touch-screen design by claiming that, "Olympus is the only company today using a touch-screen interface on a music player...." Yea, and would you like to know why? Because it's stupid.

Your Digital Sippy Cup

Sure, you've got your PSP, but what do your kids have for strutting their digital stuff? One possible choice is the Mattel Juice Box™ (Fig. 11-5). Based on an embedded Linux design made by Emsoft, the Juice Box is a 2.75-inch, color LCD music, photo, and video player (Fig. 11-6). All is not bliss in the Juice Box world, however. A special Juice Box MMC/SD media card reader program and USB card reader is needed for getting MP3 audio and still images onto the Juice Box. And, even with this special software/hardware option (marketed as the Juice Box MP3 Starter Kit), you still can't load your own video onto the Juice Box. Be advised, though, both the sound and video quality of the Juice Box are subpar. In fact, video delays and skips are par for this device.

Alternatively, you can purchase Mattel videos. Known as Juiceware™, these videos are supplied on a removable flash media card. By using the optional Juice Box MP3 Starter Kit you can rip CDs (using CoffeeCup® software) and digital pix (converted into Juice Box Picture format or .jbp) into your own personal MMC/SD juice ware, but you can't encode and transfer videos to the Juice Box.

FIGURE 11-5 **Mattel Juice Box.**

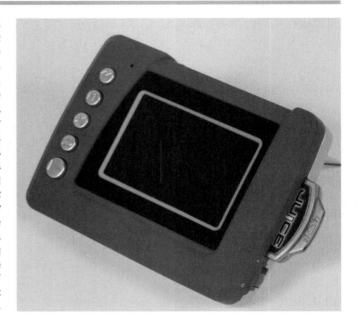

FIGURE 11-6 You can fold the plastic cover on the Juice Box backwards as a viewing stand. Even though the Juice Box is a less than a stellar performer, you can try to hack a SD/MMC media card for storing your own music and photos by following the instructions on the eLinux.org Wiki Web site (www.elinux.org/wiki/JuiceBox MMCHack).

If you'd like to try your hand at hacking the Juice Box, make sure to begin your research at www.elinux.org. Tim Riker has established a fledgling project for exploring, programming, and hacking commercial embedded Linux toys such as the Juice Box. Although Juice Box might not be the perfect mobile multimedia player for your kids, Mattel has ensured that at least your kids won't go (digitally) thirsty.

Build Proprietary Formats and They Won't Come

Attempting to hook kids on proprietary formats early, Tiger Electronics, a division of Hasbro, Inc., released the VideoNow Color in mid-2004. Derived from the mildly popular monochrome version of its debut player, VideoNow, the newer VideoNow Color sported a larger, backlit LCD-screen, slightly improved video playback, and was available in a LifeSavers assortment of colorful plastic.

Even these improvements didn't shake the fact that VideoNow Color would only play proprietary Personal Video Disc (PVD) format media. How many families could afford to have separate CD, DVD, UMD, and PVD libraries? Although the novel, original VideoNow was able to sell approximately 1.2 million units in 2003, by the middle of 2005 the novelty had worn off, and the newer VideoNow Color was relegated to store clearance shelves with retail prices slashed in half.

Chintzy Doesn't Mean Junky

Wouldn't it be great if you could make your iPod act like some sort of, I don't know, media reader, or something? Then, you could transfer, say, photographs between the iPod and your PSP. Well, you can with a very simple device—Belkin Media Reader for iPod. But the Belkin Media Reader will actually go you one better. You can use it to transfer images from other media such as CompactFlash®, SmartMedia™, and Secure Digital/MultiMediaCard, too.

Now, the Belkin Media Reader isn't really that impressive looking of a device. There is a simple iPod docking connector plug for hooking this device up to your Pod and a 5-in-1 media card port for receiving your Memory Stick PRO Duo, as well as the other supported media.

In operation, you just slap the four supplied batteries into the Belkin Media Reader, insert the Memory Stick PRO Duo, and connect it to your iPod's docking connector. That's it. Your iPod will automatically detect the photos on the device and enable you to transfer files from the PSP media card onto the iPod. Nothing could be simpler, right? Well, there are a couple of caveats that you must obey.

First, the Belkin Media Reader will only work with iPods that are running the iPod Firmware Version 2.2 and higher. Second, the photo transfer with the Belkin Media Reader is a one-way street. You can only import photos from the PSP media card onto the iPod. Not vice versa. Finally, and this is the kicker, your PSP media card must be inserted into a Memory Stick Duo Adapter such as the Sony Memory Stick Duo Adaptor ([sic]; MSAC-M2). Oh, and make sure that you heed my Power User Mega-Hack from Chapter 1: Hold all of your photos inside the special Digital Camera Images folder (e.g., DCIM/1000LYMP) on your Memory Stick PRO Duo. Follow these three little rules and you'll be in Belkin bliss.

HOW TO USE THE BELKIN MEDIA READER

BS-1 The Belkin Media Reader can copy old Memory Stick media card image files onto an Apple iPod.

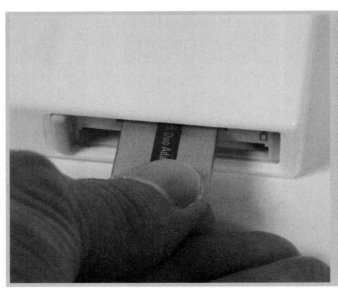

BS-2 A Sony Memory Stick Adaptor [sic] will enable you to copy image files on your Memory Stick PRO Duo media cards onto an iPod.

BS-3 There are no buttons to turn on for the Belkin Media Reader; just plug it into the iPod and it works. Isn't that nice, for once?

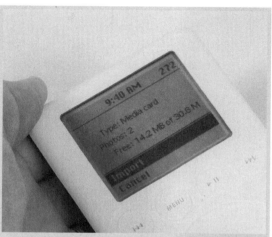

BS-4 If your iPod has Firmware 2.2, then the Belkin Media Reader will find your images on your media card. That is, provided that you followed my Power User Mega-Hack from Chapter 1.

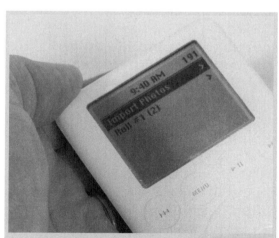

BS-5 To copy photographs from your PSP Memory Stick PRO Duo media card onto an iPod, just select "Import Photos."

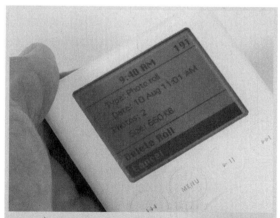

BS-6 The imported photographs are organized into a roll and stored on the iPod. Later, on your computer, you can copy these images from the iPod onto your computer.

Paper Your Planet

Ya know, it's kinda funny. Here we live in a totally digital world, yet we still rely on a lot of paper to prove our points. Take digital photography, for example. Yes, you can easily turn the digital camera around toward your friends and show them all of your wonderful candid camera images, but they just aren't that impressed looking at their antics on a small 1- to 2-inch TFT LCD, are they? Rather, these critics want to see and hold their mug shots in their little mitts. A picture is still worth about one thousand words, eh?

So consider this scenario: You just showed a spectacular slideshow of images on your PSP to a group of friends and someone says, "Hey can I have one of those pictures?" You say, "Sure, I'll Wi-Fi it to you," not realizing that "Wi-Fi" is a noun and not a verb. "No, man, I mean a 'picture.'" Gulp; you swallow hard and stutter, "I don't know how to do that."

Well, now you do. Just take your Memory Stick PRO Duo out, slap it in your printer, press a couple of buttons, and out comes the perfect print. It's that easy. What? You don't know of a printer that can do that? The Sony DPP-EX50 is just such a printer. This is an inexpensive, digital, three-pass dye sublimation printer that can spit out a print in just over one minute. And, unlike the Belkin Media Reader, the Sony DPP-EX50 does not require that your photographs be sequestered inside the DCIM/100OLYMP folder hierarchy. You can print them directly from the PSP/PHOTO folder.

Just like the Belkin Media Reader, however, you *must* use a Memory Stick Duo Adapter such as the Sony Memory Stick Duo Adaptor ([sic]; MSAC-M2) for adapting your PSP media card to the Sony DPP-EX50. Once you've got your media card in the printer, the photos folder will be found, and the name of the first image will be displayed on the printer's LCD screen. So, make sure that you use descriptive names for your photos. Alternatively, you can print an index print that will combine all of your media card's photos onto a single print. Then, select the thumbnail image that you want to make into a full-size print, press the arrow button to display that image's filename on the DPP-EX50's LCD screen, and press the Print button.

In less time than it takes to read this page, you will have a beautiful print that is an equal to the exquisite visual output that is generated by your PSP's LCD screen. Seeing is believing, and having a photographic print in your hand is a fantastic extension to the list of capabilities that are espoused by the PSP. No, the DPP-EX50 is not a portable digital printer—it requires an AC power supply. And no, the DPP-EX50 is not lightweight—it tilts the scale at approximately 4 $1/2$ pounds. But if you want fast, easy, beautiful prints that match the LCD screen output from your PSP, then the Sony DPP-EX50 will give you that capability.

HOW TO PRINT PHOTOS FROM YOUR PSP

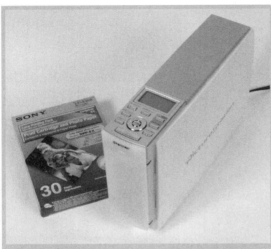

PS-1 The Sony DPP-EX50 is a great printer for making prints of your fav PSP photos.

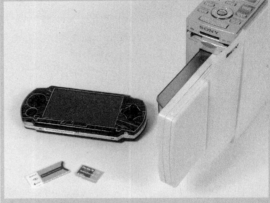

PS-2 Your PSP Memory Stick PRO Duo media cards won't work in the DPP-EX50 digital printer. You will need either a Sony Memory Stick Adaptor [sic] or the ability to copy your images to an older Memory Stick.

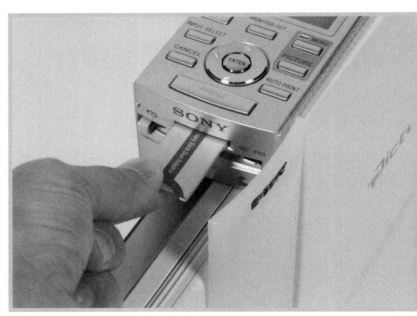

PS-3 Insert the adapter and the DPP-EX50 will automatically locate your digital images. A special note: this feature will work with both the default PSP PHOTO folder hierarchy and my Power User Mega-Hack in Chapter 1 for loading your PSP images inside a special DCIM folder. The choice of where you keep your pix is yours. But for maximum image flexibility, I would recommend using only the DCIM technique for all PSP images.

PS-4 Don't know which print you want? Print an index of all your images first. Then you can scroll to the correct image and make a print.

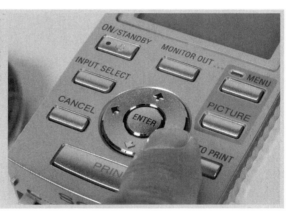

PS-5 A simple arrow key button allows you to scroll through all of your images until you find the one that you want to print.

PS-6 A print straight from your PSP in less than 2 minutes.

CHAPTER 12
PSP P.S.

What does the future hold in store for the PSP? I don't know for sure, but one thing is certain: The PlayStation®3 (PS3) will be an integral part of the complete PSP system.

Hard-core gamers can probably still remember where they were when they first learned about the PS3. For me, the venue for my revelation was an event known as E3.

The E3 is the premiere event for seeing every significant upcoming interactive gaming product all jointly displayed inside the same building. This is the perfect site for gaining access to advanced game industry information, press releases, and prototypes.

Prior to the start of E3 2005, on May 16, 2005, SCEI held a press conference to unveil just such a prototype—the PS3 (Fig. 12-1). In marked contrast to the diminutive and sealed PS2 enclosure, the PS3 was powerful, accessible, attractive, and integrated with the PSP.

Esprit de Core

From the get-go, the PS3 is a gamer to its core. A 3.2-GHz PowerPC core, that is. Organized into a revolutionary processor called "Cell," the PowerPC core was jointly developed by IBM, Sony Group, and Toshiba Corporation. Blazing along at processing speeds of nearly 2 teraflops, the Cell processor is able to drive

FIGURE 12-1
The Sony PlayStation 3. (Based on prototypes photographed at E3 2005.)

an RSX graphics processor that was codeveloped by NVIDIA Corporation and SCEI along with 256 MB of XDR main RAM developed by Rambus, Inc., and 256 MB of GDDR3 VRAM.

The Cell processor isn't the only new technology inside the PS3. A 54-GB (in dual-layer format) Blu-Ray Disc ROM (BD- ROM) will supply full high-definition-quality content. That isn't to say that SCEI will leave all of your PS2 library out in the cold. Rather, PS3 will be able to read CD-ROM, DVD-ROM, DualDisc, and SACD, as well.

All of the other usual interface suspects will be in the PS3, too. Expect USB, Memory Stick media cards, Bluetooth, Ethernet, and Wi-Fi to be accessible on the PS3. Better yet, a detachable 2.5-inch hard disk drive can be slapped into a special storage slot. But the best interface news for handheld owners is that the PS3 and PSP will be able to, in the words of an SCEI rep, "interact" with each other via Wi-Fi networking.

Regardless of how the next generation of video game console wars play out, neither Xbox 360 nor Nintendo Revolution are poised to become a valid digital lifestyle expression without a device such as the PSP handling the mobile aspects of our daily lives. Both Nintendo DS and Game Boy Advance SP are weak media players that are pale impressions of the PSP. Likewise, Microsoft's best effort at a mobile entertainment system would be equivalent to the ill-conceived Soviet-era-design Tablet PC.

Early release dates don't make success stories, only daily headlines. Watch for the manufacturer who integrates your new PSP passion into a true digital lifestyle paradigm. Then get ready to shift your life into high (definition) gear. Me, I'm betting on SCEI.

Epilogue

A continual raging debate circles around video games regarding the validity of health risk claims that can be directly attributed to using this type of electronic entertainment. More times than not, these pundits are poking their inquisitive probes into action-type games that rely on violent or criminal interactions between the player and the game. Think *Grand Theft Auto: San Andreas* with its "Hot Coffee" hack, rather than, say, *PONG*.

Although I will continually disagree with these types of claims (heck, they cited *The Catcher in the Rye* as "contributing to the delinquency of our nation's youth" when I was a youth), I finally ran across a news report that might be the first truly credible authentic account of the harmful health effects that can be attributable to video games. This is a tragic tale that was reported by Reuters on August 9, 2005.

You couldn't miss the headline for this story: "Man Dies after Online Game Marathon." Whoa, that's a good one. Unfortunately, the facts included with this news story were very thin, to say the least. Likewise, several of the story's key facts were downright hilarious. Unintentional, to be sure. But ridiculous, nonetheless.

In case you missed this story, here is a brief summary from this news report:

"A South Korean man who played computer games for 50 hours almost non-stop died of heart failure minutes after finishing his mammoth session in an Internet cafe...."

DP: What games?

"He only left the spot over the next three days to go to the toilet and take brief naps on a makeshift bed...."

DP: You're kidding; what cafe has a bed?

"The 28-year-old man, identified only by his family name Lee...."

DP: That's it? "Lee"? No first name?

"Lee had recently quit his job to spend more time playing games...."

DP: You're kidding; what games?

"South Korea, one of the most wired countries in the world, has a large and highly developed game industry."

DP: Yea, like "killer games."

FIGURE E-1 **Don't be a Lee—take it out.**

So will video games kill ya? I don't know, but don't be a Lee—take it out (Fig. E-1).

Start a fashionable video gaming revolution—go PSP.

APPENDIX A
PS2Pdp

How to Make a PS2 Portable Game Machine

Do you ever look at your PSP and say, with an accompanying head slap, "Gee, I wish that I would've thought of that." In other words, you try to convince yourself that *you* could have invented a portable PlayStation game machine. Oh, yes; given enough time, money, and talent. Well, this might not be as ridiculous of an idea as you first thought. Is your head still hurting?

Just look how simple this design plan is:

→ Portable power supply
→ Portable display screen
→ Powerful game console

Easy, right? OK, let's get to work. First of all, let's focus our attention on defining the actual components that will constitute our three design ingredients:

→ Portable power supply—a portable DVD rechargeable battery pack that is capable of supplying 8.5 volts
→ Portable display screen—a reasonable LCD screen with audio/video (A/V) inputs, on-board speakers, and self-contained power supply

A-1 If you're having trouble locating a suitable battery pack, then consider the model BG 9-12-60 from batterygeek.net. Otherwise, you can roll your own with conventional battery packs and a suitable power-supply plug (e.g., Digi-Key part number CP-2194-ND).

A-2 Connect the Sony-supplied A/V output plug to your PS2.

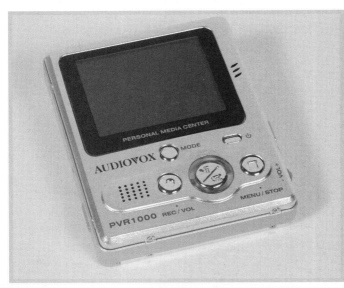

A-3 Find a good, reliable LCD screen that has A/V *inputs*. Most portable DVD players do *not* have inputs. So, be careful what brand you choose. Also, watch your power supply. A portable device with its own rechargeable power supply, such as this Audiovox PVR1000, is ideal for this purpose.

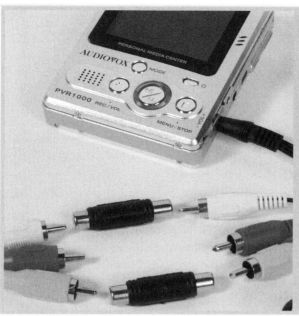

A-4 **Do the hookup snake dance.** You'll probably need three RCA phono male-to-male adapter plugs. You can locate these adapters at RadioShack. (One of the adapter plugs has been temporarily eliminated from this photo.)

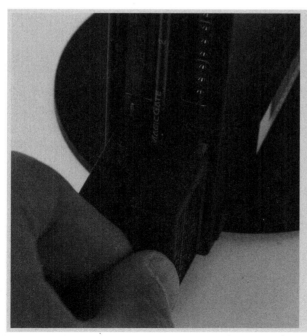

A-5 **Plug in a DualShock2 game controller and test your PS2Pdp someplace really remote...such as the kitchen table during mealtime.

→ Powerful game console—a system with a small footprint, a big game library, and the ability to play DVD movies; the Sony slim PS2 is the best choice here

Before we begin, we must pay homage to the first known hacker to build a PS2 portable game machine—Benjamin J. Heckendorn. While most of us were still in diapers, Heckendorn was able to cram his entire PS2 console, battery, and LCD-screen assembly into a single, portable, gamer's *pièce de résistance*. He called his creation the "PS2p," a very impressive project and worthy of closer study (www.benheck.com).

Even though I'm still wearing diapers, I thought that I could make a simpler version of Heckendorn's PS2p that wouldn't require much soldering, could be transported in an automobile, and could sit comfortably on my oldest daughter's dresser (for, ahem, game review purposes only). I would call my creation the "PS2Pdp." Cute, eh? That final "dp" bit, I mean. Well, anyway, if you currently have one of the newer slim PS2 consoles, a suitable portable display that has A/V inputs, and the desire to experiment with hooking an external DVD battery up to your console, then this hack is for you. If not, well, then please turn to the next appendix, weenie.

APPENDIX B

Ten Games Ya Gotta Play

How do you find the tastiest game offerings out of a tidal wave of PSP titles? Well, if you have deep enough pockets, you could spend about $5000 and try them all. Unfortunately, you would be left with a fairly sizable pile of UMD garbage. A more economical and logical notion would be to study the following short list of the 10 best PSP games (Fig. B-1). Then refine your

FIGURE B-1
There are some great PSP games out there. The problem is trying to separate the good ones from the bad ones. Thankfully, this appendix does that for you.

choice to the genres that interest you most, make out your wish list, and start playing. One game at a time.

This list is arranged in alphabetical order according to game title.

TITLE: *Archer Maclean's Mercury*
DISTRIBUTOR: Ignition Entertainment Ltd.
GENRE: Puzzle
RATING: E (Everyone)
OPTIONS: Wi-Fi, 1 or 2 players; 2 MB of Memory Stick PRO Duo
THE BOTTOM LINE: How hard can it be to roll a bubble of liquid mercury around on your lap? Very hard.

FIGURE B-2 Unlike other titles in the *Armored Core* franchise, *Armored Core: Formula Front—Special Edition* employs an artificial intelligence (AI) component that elevates strategy game play to the same level as the game's omnipresent action elements. (Photograph courtesy of Agetec, Inc.)

TITLE: *Armored Core: Formula Front—Special Edition* (Fig. B-2)
DISTRIBUTOR: Agetec, Inc.
GENRE: Action/Strategy
RATING: Rating pending; could be T (Teen)
OPTIONS: Wi-Fi, 1 or 2 players; connectivity to PS2
THE BOTTOM LINE: AI, AI, it's off to annihilate we go.

TITLE: *Frogger Helmet Chaos*
DISTRIBUTOR: Konami Digital Entertainment—America
GENRE: Action/Adventure

RATING: E (Everyone)
OPTIONS: Wi-Fi, 1 to 4 players; bonus—original arcade game locked inside...somewhere
THE BOTTOM LINE: A classic that sticks with you like roadkill on the highway.

TITLE: *Ghost in the Shell®—Stand Alone Complex™*
DISTRIBUTOR: Bandai America
GENRE: Action; role-playing game
RATING: T (Teen)
OPTIONS: Wi-Fi; bonus features
THE BOTTOM LINE: Crack the code to bust the cells.

TITLE: *Gretzky NHL*
DISTRIBUTOR: SCEA
GENRE: Sport
RATING: E (Everyone)
OPTIONS: Wi-Fi, 1 or 2 players
THE BOTTOM LINE: The coolest sports game that you can hold in your lap.

TITLE: *Infected*
DISTRIBUTOR: Majesco Entertainment
GENRE: Third-person shooter
RATING: Rating pending; could be T (Teen) or M (Mature)
OPTIONS: Wi-Fi
THE BOTTOM LINE: Give the gift of sharing—a drop of immune blood fired from a high-velocity viral gun.

FIGURE B-3 Prince Pietro journeys through enchanted lands battling dragons while filling your PSP up with some tasty Japanese animation—anime. (Photograph courtesy of Agetec, Inc.)

TITLE: *PoPoLoCrois* (Fig. B-3)
DISTRIBUTOR: Agetec, Inc.
GENRE: Role-playing game
RATING: Rating pending; could be E (Everyone)
OPTIONS: Bonus—anime sequences from the TV series
THE BOTTOM LINE: PoPoSoCool.

TITLE: *Rengoku™: The Tower of Purgatory*
DISTRIBUTOR: Konami Digital Entertainment—America
GENRE: Action
RATING: M (Mature; blood and violence)
OPTIONS: Wi-Fi, 1 to 4 players; 1 MB of Memory Stick PRO Duo
THE BOTTOM LINE: This Adam ain't ever going to find Eden.

TITLE: *Spider-Man 2*
DISTRIBUTOR: Activision, Inc.
GENRE: Action
RATING: T (Teen; violence)
OPTIONS: 155 KB of Memory Stick PRO Duo
THE BOTTOM LINE: Spidey high; Spidey go.

TITLE: *Ys: The Ark of Napishtim*
DISTRIBUTOR: Konami Digital Entertainment—America
GENRE: Role-playing game
RATING: Rating pending; could be T (Teen) or M (Mature)
OPTIONS: Bonus mini-games
THE BOTTOM LINE: Just make sure that you don't Ark up the wrong tree.

The game *Theseis* by track7games also makes this list, even though no definite release date could be guaranteed prior to publication. This exclusive advance game information was provided to me by the good folks at track7games:

TITLE: *Theseis: Journey into Hades to Unveil the Mystery* (Fig. B-4)
GENRE: Adventure/Action
RATING: Rating pending; could be T (Teen) or E (Everyone)
THE BOTTOM LINE: You may go against the current, but just don't skinny dip in the River Styx.

FIGURE B-4
Groundbreaking graphics and mind-numbing mystery are hallmarks of the revolutionary game *Theseis* by track7games. (Photograph courtesy of track7games)

APPENDIX C
Ten Movies Ya Gotta Watch

The big screen isn't the same as the small-but-wide screen of the PSP. Movies that are basically awful (like *Resident Evil: Apocalypse* and *Sisterhood*) might seem impressive on a theater screen, only to become worthless wastes of time and money on the PSP. Why the difference, you ask? Well, my guess is that the PSP is a very personal movie theater that enables you

FIGURE C-1
A good flick is more than just a subjective matter of taste, bud. The PSP acts like a motion picture microscope that can help you spy cinematic brilliance, as well as amplify doggie boners.

to more closely scrutinize your video content. Most sane people are able to spot crap fairly easily if given a chance. And the PSP will give you that chance.

More than 50 UMD movies were studied during the preparation of the following list (Fig. C-1), which is arranged in alphabetical order according to movie title.

TITLE: *Are We There Yet?*
STUDIO: Revolution Studios
RATING: PG; language, rude humor
THE BOTTOM LINE: Planes, trains, and a Lincoln Navigator, cubed.

TITLE: *Hellboy: Director's Cut*
STUDIO: Revolution Studios
RATING: Unrated; should be PG-13; violence and language
THE BOTTOM LINE: Basically, I would just keep the horns; they help to soften your oval face.

TITLE: *I, Robot*
STUDIO: Twentieth Century Fox Home Entertainment
RATING: PG-13; intense, stylized action and brief, don't-blink nudity
THE BOTTOM LINE: What's with all of that sweet potato pie, Sonny?

TITLE: *Kill Bill, Volume 1 & Volume 2*
STUDIO: Miramax Films
RATING: R; strong language, bloody violence, and some sexual content
THE BOTTOM LINE: Gore noir with gored sword chord.

TITLE: *Magic of Flight* (IMAX)
STUDIO: Image Entertainment
RATING: Unrated; should be G; general audiences
THE BOTTOM LINE: It's just like flying, except you're not.

TITLE: *Pirates of the Caribbean: The Curse of the Black Pearl*
STUDIO: Walt Disney Pictures
RATING: PG-13; action/adventure violence

THE BOTTOM LINE: You'd have to sail more than seven seas to ever meet an effeminate zombie pirate.

TITLE: *Reign of Fire*
STUDIO: Touchstone Pictures
RATING: PG-13; intense action and violence
THE BOTTOM LINE: These pigeons do more than just poop on statues.

FIGURE C-2 *Samurai Champloo* includes some cutting-edge fight scenes. [Photograph courtesy of Geneon Entertainment (USA)]

TITLE: *Samurai Champloo, Episodes 1 & 2* (Fig. C-2)
STUDIO: Geneon Entertainment (USA), Inc., for Fuji Television
RATING: "Suggested 16 & up"; could be R; violence and strong language
THE BOTTOM LINE: Hi-ya; sunflowers don't have an odor, do they?
SPECIAL NOTE: There are four other highly recommended anime titles from Geneon: *Akira* (this classic 1988 film has been digitally re-mastered for the PSP), *Appleseed* (a visual treat with a surprise Ryuichi Sakamoto soundtrack contribution; Figs. C-3 and C-4), *Hellsing* (three episodes from this violent anime; Fig. C-5), and *Gungrave* (three episodes of hip violence; Fig. C-6).

FIGURE C-3 *Appleseed* is a mature anime movie that sounds as good as it looks. [Photograph courtesy of Geneon Entertainment (USA)]

FIGURE C-4 This cityscape from *Appleseed* is of Earth's last city, Olympus. [Photograph courtesy of Geneon Entertainment (USA)]

FIGURE C-5 In *Hellsing*, the undead actually go clubbing. [Photograph courtesy of Geneon Entertainment (USA)]

FIGURE C-6 *Gungrave* is vengeance from beyond the grave, with guns. Go figure. [Photograph courtesy of Geneon Entertainment (USA)]

TITLE: *Saw*
STUDIO: Twisted Pictures
RATING: R; strong language, grisly violence
THE BOTTOM LINE: Hey, that's why you have two feet.

TITLE: *Steve Oedekerk Presents: Thumb Wars and Thumbtanic*
STUDIO: Image Entertainment, Inc.
RATING: Not rated; should be PG; intense nail clipping
THE BOTTOM LINE: A real nail biter coupled with a white knuckler.

APPENDIX D
Things to Know, Places to Go

Here is a collection of every reference, document, and Web site that is mentioned in this book. Live and learn.

3GP Converter 031—www.nurs.or.jp/~calcium/3gpp/3GP_Converter031.zip
Adobe Studio Exchange—http://share.studio.adobe.com/default.asp
Aiptek—www.aiptek.com
Apple Computer—www.apple.com
ARM Station—www.armstation.com
AS&C CooLight—www.coolight.com
BatteryGeek Inc.—www.batterygeek.net
Beal Systems Glowire—www.glowire.com
Belkin Corporation—www.belkin.com/index.asp
bhv Software GmbH & Co. KG—www.bhv.de
BoxWave Corporation—www.boxwave.com
Cowon Systems, Inc.—http://eng.cowon.com/index.php
Digi-Key Corporation—www.digikey.com

Easybuy2000.com—
www.easybuy2000.com/store/nintendo%20accessories/index.htm

elinux.org—http://elinux.org/wiki/JuiceBox

Elwirecheap.com—www.elwirecheap.com

Emsoft—www.emsoftltd.com

Entertainment Software Association—www.theesa.com/index.php

The 2005 Entertainment Software Association's Essential Facts about the Computer and Video Game Industry—
www.theesa.com/files/2005EssentialFacts.pdf

Everything USB—www.everythingusb.com

eXpansys—www.expansys-usa.com

ffmegX—http://homepage.mac.com/major4

Flash Linker & Card Set—
http://gameboy-advance.net/flash_card/gba_X-ROM.htm

Future Technology Devices International Ltd. (FTDI)—
www.ftdichip.com/Resources/Utilities.htm

Game Boy Advance E-Book—
http://members.optushome.com.au/dancotter/ebook.htm

Game Boy Advance ROMs—www.gameboy-advance-roms.info;
www.gameboy-advance-roms.com; www.gba-rom.com

Game Boy Games—www.gameboy-games.com

GameDAILY Biz: The Game Industry Professional's News Source—
http://biz.gamedaily.com/articles.asp?section=news

Gamer Graffix—www.gamergraffix.com

GBA X-ROM 512-Mb Flash Linker and Card Set—
http://gameboy-advance.net/flash_card/gba_X-ROM.htm

"Hackers Add Web, Chat to PSP," April 5, 2005—www.CNN.com
(exact link information is unavailable)

Hacking Video Game Consoles by Benjamin J. Heckendorn
(John Wiley & Sons, 2005)

HomebrewPSP Converter—http://ipsp.kaisakura.com/homebrew.php

International Game Developers Association (IGDA)—www.igda.org

Juice Box—www.juicebox.com/home.aspx

Juice Box SD/MMC Hack—www.elinux.org/wiki/JuiceBoxMMCHack

Junxion Box—www.junxionbox.com

KXploit (direct loader) version 1.50 by PSP-Dev Team—
http://psp-dev.1emulation.com

lantus—www.lantus-x.com/PSP

Linksys—www.linksys.com

Lua Player—www.luaplayer.org

Lua programming language—www.lua.org

Mac OS 10.4 technical document
"How to prevent .DS_Store file creation over network connections"—
http://docs.info.apple.com/article.html?artnum=301711

Made in Japan by Akio Morita with Edwin M. Reingold and
Mitsuko Shimomura (E. P. Dutton, 1986)

MAKE magazine—www.makezine.com

Mattel Juice Box—www.juicebox.com/home.aspx

"The Memory Stick" by Shigeo Araki (*IEEE Micro,* Vol. 20, No. 4, 2000)—
http://csdl2.computer.org/persagen/DLAbsToc.jsp?resourcePath=/dl/mags/
mi/&toc=comp/mags/mi/2000/04/m4toc.xml&DOI=10.1109/40.865865

Memory Stick Business Center—www.memorystick.com

Norêve—www.noreve.com

NPD Funworld—sales and consumer tracking of toys and video games
industry—www.npdfunworld.com/funServlet?nextpage=index.html

The NPD Group—global sales and marketing information—www.npd.com

Official Web site of Benjamin J. Heckendorn—www.benheck.com

Official WipEout PurE Web site—www.wipeoutpure.com

Olympus Imaging America Inc.—www.olympusamerica.com

Otterbox—www.otterbox.com

Pelican—www.pelicanperformance.com

Phoenix: The Rise and Fall of Videogames, 3d ed., by Leonard Herman
(Rolenta Press, Springfield, N.J., 2001)—www.rolentapress.com

PS2NFO—www.ps2nfo.com

PSP-Dev—http://psp-dev.1emulation.com

PSP emulators—www.dcemu.co.uk/vbulletin/showthread.php?t=7218

PSP hacks—www.psp-hacks.com

PSP updates—www.pspupdates.com

PSP Video 9—www.pspvideo9.com

PSPE PSP emulator for Windows—http://psp-news.dcemu.co.uk/pspe.htm

PSPWare for Mac OS X—www.nullriver.com/index/products/pspware

RnSK Softronics—http://ipsp.kaisakura.com/index.php

Sharp LQ038Q5DR01 3.8-inch, color TFT LCD module—www.sharpmeg.com/part.php?PartID=628

Sharp Microelectronics products—LCDs—www.sharpmeg.com/productgroup.php?ProductGroupID=51

Simon Sheppard's Mac OS X command line command list—www.ss64.com/osx/

Sony Ericsson—www.sonyericsson.com

StuffIt Deluxe—www.stuffit.com/mac/index.html

Torche Elwire—www.elwire.com

Ultrahigh-capacity external, portable DVD battery model number BG 9-12-60—www.batterygeek.net/BG91260.htm

USB Implementers Forum, Inc.—www.usb.org

USBMan—www.usbman.com

Vaja—www.vaja.biz

Vaja Cases—www.vajacases.com

Vaja Cases Choice custom order—www.vajachoice.com

VideoNow Color—www.hasbro.com/videonow

Wi-Fi Alliance—www.wi-fi.org

Wi-Fi-FreeSpot™ directory—www.wififreespot.com

WiFi Seeker—www.wifiseeker.com

Wi-FiHotSpotList.com—www.wi-fihotspotlist.com

WinRAR, RAR, and ZIP archive tool—www.rarlab.com/download.htm

X-OOM Movies on PSP—www.x-oom.com

INDEX

A2, 184
Accessing Web portals, 51–55
Accessories (see PSP accessories)
ActionGrip, 139
Ad hoc WLAN, 46, 47
Adobe Photoshop, 30, 41, 136
Adobe Studio Exchange Web site, 136
AIBO, 5
Aiptek IS-DV, 169, 173–176
Airport Extreme Base Station, 47, 48, 147, 150
Akira, 211
Amazon.com, 162
Ambition (leather), 133
Apparel market, 5
Apple Airport Extreme Base Station, 47, 48, 147, 150
Appleseed, 211, 212
Archer Maclean's Mercury, 89, 204
Are We There Yet?, 210
A.R.M. Kit, 76
A.R.M. Kit Memory Stick Jacket for P800, 76, 78, 79
A.R.M. Kit-Screen Protector, 80
A.R.M. Memory Stick Jacket, 76
ARM Station, 76
Armored Core: Formula Front— Special Edition, 115, 204
AS&C CooLight, 108
ASF tracks, 170, 175–177
Atari Lynx, 114
Audiovox Corporation, 179
Audiovox PVR1000, 170, 179–181, 200
Author (Prochnow), 127, 198
AVI files, 175, 177

Battery, charging the, 22–23, 25, 159–162
"Battery Information" screen, 22, 24
batterygeek.net, 200
BBEdit, 123
Beal Systems GLOWIRE, 108
Belkin Media Reader, 187–189
Bengtsson, Joachim, 123

Betamax, 5
BG 9-12-60, 200
Boxwave Corporation, 106, 161
Boxwave Corporation PSP ActionGrip, 139
Brightness, 23, 37
Buss, Frank, 123

Car charger, 19
Cell phone, 74
Cell processor, 195–196
Charge connector, 19
Charging the battery, 22–23, 25
Clothing (apparel) market, 5–6
CoffeeCup, 185
Columbia Pictures Entertainment, Inc., 5
Comfort ear gel plugs, 100
Contact information, 215–218
Copy/convert video, 37, 40–45
Cowon A2 portable multimedia player, 184
Cowon American iAUDIO X5, 162–171, 181–184
Cowon Systems, 184
Crystal View Edition (screen protector), 80
Custom skin, 140–143

DailySoft, 76
DCIM folder, 31–32
Digi-Key Corporation, 57
Digital camera, 159
Digital Camera Images, 32
Digital image editing, 30–32
Disassembling the PSP, 57–71
Disclaimer, 60
Discman, 5
Display Button, 23
DOOM, 119, 120
DPP-EX50, 190–193
Dr. Bott ExtendAir Omni Antenna, 150
Dragon images, 136
.DS_Store, 33
DualShock2 game controller, 201
DVD-AR, 25

DVD-Audio, 24
DVD-R, 24
DVD-R/RW, 24
DVD+R/RW, 25
DVD-RAM, 25
DVD-ROM, 24
DVD+RW, 25
DVD-SR, 25
DVD-Video, 24
DVD-VR, 25

E3 2005, 195
Ear bud headphones, 100
Easybuy2000.com, 115
EL wire mod, 105–109
Electric rice cooker, 5
Electroluminescent (EL) wire mod, 105–109
Elwirecheap.com, 108
Emsoft, 185
Emulating other game systems, 119–122
Ethernet, 41
eXpansys, 76

FAQs (see Questions and answers)
ffmegX, 44
File names, 32
Firm upgrades, 25–26, 90
Firmware 1.50, 10, 121
Firmware 1.51, 10
Firmware 2.0, 10, 55
512-Mb Flash ROM cartridge, 116
Flash Linker & Card Set, 115–118
Flash ROM cartridge, 116
FlexiSkin, 106, 107
Fold-down subwoofer unit, 97
Folder, 26
Folder hierarchy, 31
Folder names, 32
French leather case, 131–133
Frogger Helmet Chaos, 204–205
FTDI Web site, 159
FX300 Racing League, 51

Game Adapter, 147, 148
Game Boy, 114

219

Game Boy Advance (GBA), 114
Game Boy Advance SP, 114
Gamer Graffix, 126–127, 132
Gamer Graffix Skinz, 126–132
Games (*see* Video games)
GBA, 114
GBA media player (GBAmp), 115–118
Geek chic, 126
Getting under the hood (disassembly), 57–71
Ghost in the Shell—Stand Alone Complex, 205
Ginza Heating Company, 2
GLAY'z, 88
GoGear camcorder, 169
Gran Turismo 4: The Real Driving Simulator, 12
Gretzky NHL, 205
Gungrave, 211, 212

"Hackers Add Web, Chat to PSP," 51
Handheld entertainment system, 8
Headphones, 98–100
Health risk, 197–198
HEAVY (sound button), 24
Heckendorn, Benjamin J., 57, 202, 217
Hellboy: Director's Cut, 210
Hellsing, 211, 212
Hirai, Kaz, 8
Homebrew games, 12–13, 118–124
Homebrew PSP Converter for Mac OS X 10.4.x, 122
House of Flying Daggers, 87
Hurricane Dennis, 147, 150

I, Robot, 210
iAUDIO X5, 162–171, 181–184
Ibuka, Masaru, 1, 3
Image Mate 12-in-1 reader/writer, 28, 29
Image viewing capability, 30–32
Incompatible Data music files, 33, 36, 38–39
index.html, 52
Infected, 205

Infrastructure WLAN, 47
Inverter, 108
Invisible Mac files, 33, 36, 38–39
iPod, 9–10, 189
iPod mini, 178–179
iPod shuffle, 167–168
IS-DV, 169, 173–176
iTunes, 178
JAZZ (sound button), 24
JetAudio, 162, 184
Jewel's Knights, 136
JPEG, 30, 41
Juice Box, 185–186
Juice Box MPC/SD media card reader program, 185
Juice Box MP3 Starter Kit, 185
Juiceware, 185
Jukebox, 170, 176
Junxion Box, 151

Kids (Juice Box), 185–186
Kill Bill, Volume 1 & Volume 2, 210
KXploit, 118, 119

lantus, 119
LCD driver and backlight connectors, 66
LCD screen, 57
LED glow buy mod, 109–111
Linksys Wireless-B Game Adapter, 147, 148
Lists:
 EL wire suppliers, 108
 games—U.S. PSP launch, 14
 product flops, 5
 products that use Memory Stick, 25
 PSP emulators, 122
 supplier, games, products, etc., 215–218
 top ten games, 203–207
 top ten movies, 209–213
 UMD movie titles, 91–94
LittleWriter, 117, 188
Logic3 PSP Sound Grip, 138, 139
Logic3 Sound system, 97, 98
Logitech PlayGear Amp, 100, 101

Logitech PlayGear Mod headphones, 98
Logitech PlayGear Pocket, 130
Logitech PlayGear Pocket case, 136
Logitech PlayGear Pocket skin, 140–143
Logitech PlayGear Share stereo adapter, 99
Logitech PlayGear Stealth headphones, 100
Lowser, 123
LQ038Q5DR01, 57
LQ038Q5DR01 KX15-40K-type connector, 60
LQ043T3DX01, 57
Lua Player, 123–124
Lua Player Web site, 124
Lua Programming Language, 123
Lynx, 114

Mac gamers, 122
Made in Japan (Morita), 3
Magic Flight (IMAX), 210
MagicGate, 27, 74
Make: Technology on Your Time, 119
"Man Dies after Online Game Marathon," 197
Mandatory PSP firmware upgrades, 25–26, 90
Mattel Electronics LED, 114
Mattel Juice Box, 185–186
Mattel videos, 185
Mavica, 5
Memory Stick, 5
 ATRAC 3 file format, 27
 formatting, 26
 removing, 30
 SanDisk media, 27–29
 space, 27
 types, 21, 27
 using, 29–30
 using older media cards, 73–80
"Memory Stick, The" (Araki), 74
Memory Stick Duo, 27
Memory Stick Duo slot cover, 78
Memory Stick Expansion Jacket, 76
Memory Stick Format setting, 26

Memory Stick function, 26
Memory Stick PRO, 27
Memory Stick PRO Duo, 27
Memory Stick PRO Duo pin outs, 77
Memory Stock Duo adapter connector, 76, 77
Mercury, 89, 204
Metal Gear Acid, 171
Metal Gear Acid 3, 171
Metal Gear Acid Solid 3, 171
MobileMate, 29
Morita, Akio, 1, 3, 4
Movie conversion, 37, 40–45
Movies, 13, 15, 24–26
  convert ASF tracks to MPEG4 movies, 175–177
  copy/convert video, 37, 40–45
  playing the movie, 26
  top ten movies, 209–213
  UMD titles, 91–94
  (*See also* UMD)
MP4 file, 40–41
MPEG4 files, 40, 43
MR-500i, 184
M3U file, 29, 47
Multiplayer games, 88
Music, 23–24
  incompatible data files, 33, 36, 38–39
  invisible Mac files, 33, 36, 38–39
  iPod mini, 178–179
  iPod shuffle, 167–168
  iTunes, 178
  Juice Box, 185–186 38–39
  jukebox, 176
  playlist, 29, 47
  stereo sound (headphones), 95–102
Music video, 80–85
MUTE (sound button), 24
M4Vxxxxx.MP4, 40
M4Vxxxxx.THM, 41

Names of suppliers, games, products, 215–218
Network Settings option, 41, 46, 47, 152
Nevyn, 123
Nintendo DS, 114

Nintendo Game Boy, 114
Nintendo Game Boy Advance (GBA), 114
Norêve, 131
Norêve PSP case, 131–133
Norman, Michael, 136
Nullriver Software, 33

OFF (sound button), 24
Olympus m:robe MR-500i, 184
1-watt stereo amplifier, 95–102

Pacman clone, 121
PasteIn command, 167
Peerless, Jim, 121
Pelican car charger, 19
Pelican face armor, 137
Pelican Skin Grip, 138
Perfume (personal fragrance) industry, 5
Personal Video Recorder PVR1000, 170, 179–181, 200
Philips GoGear Wearable Digital Camcorder, 169
Phillips-head screw, 57
Phoenix: The Rise and Fall of Videogames (Herman), 113
PHOTO folder, 35
Photographs:
  Belkin Media Reader, 187–189
  GoGear camcorder, 169
  Juice Box, 185–186
  print (DPP-EX50), 190–193
  PSP, 30–32, 34–35
Photoshop, 30, 41, 136
Pin outs:
  LQ038Q5DR01 connector, 60
  Memory Stick PRO Duo, 77
  USB, 158
Pirates of the Caribbean: The Curse of the Black Pearl, 210–211
PlayGear Amp, 100, 101
PlayGear Mod headphones, 98
PlayGear Pocket, 136
PlayGear Pocket skin, 140–143
PlayGear Share adapter, 99
Playlist, 29, 47
PlayStation, 4, 7
PlayStation 2 (PS2), 6, 145, 147, 148–149

PlayStation 3 (PS3), 195–196
PlayStation game library, 12
PlayStation Portable (PSP):
  accessories, 10–12
  (*See also* PSP accessories)
  battery, 22–23, 25
  charge connector, 19
  charging the battery, 22–23, 25
  components, 18–21
  disassembling, 57–71
  firmware, 10
  initial PSP Value Pack, 13
  marketing campaign, 10–11
  name, 8
  original U.S. model, 8
  power switch, 18
  Prêt à PSP, 10–11
  product launch, 4–5, 8–11, 14
  remote control, 48–50
  scripting language, 123–124
  UMD disk drive cover, 20
  UMD release switch, 20
  USB port, 18
  used machine, 60
  volume, 25
  Wi-Fi switch, 18
  WLAN switch, 46
Polishing cloth, 125–126, 131
PoPoLoCrois, 119, 205
POPS (sound button), 24
Portable entertainment system, 8
Portable game machines (*see* Video games)
Pouch, 125
Power charger plug, 67
Power Director, 176, 177
Power switch, 18
Power User Mega-Hack, 31–32
Prêt à PSP, 10–11
Printer (DPP-EX50), 190–193
Prochnow, Dave, 127, 198
Product failures, 5
Product launch, 4–5, 8–11, 14
PS2, 6, 145, 147, 148–149
PS2 DVD-ROM, 25
PS2 portable game machine (PS2Pdp), 199–202
PS2 SCPH-70000 series, 6
PS2p, 202

PS2Pdp, 199–202
PS3, 195–196
PSP (see PlayStation Portable)
PSP + PS2 (*Metal Gear Acid*), 171
PSP accessories, 10–12, 125
  custom skin, 140–143
  PSP case, 131–139
  skin, 126–132, 140–143
  Value Pack accessories, 125–126
PSP ActionGrip, 106, 107
PSP case, 131–139
PSP emulator for Windows (PSPE), 124
PSP emulators, 122
PSP fob, 130
PSP folder hierarchy, 31
PSP games (see Video games)
PSP image viewing capability, 30–32
PSP LCD, 57
PSP movies (see Movies)
PSP remote control, 48–50
PSP skin, 126–132
PSP Value Pack, 13
PSP Value Pack accessories, 125
PSP *Wipeout Pure* hack, 51–55
PSPacman, 121
PSPcast, 80–85
PSPE, 124
PSPTV, 80–85
PSPWare, 40, 44, 45
PSPWare for Mac OS X, 33
PVR1000, 170, 179–181, 200

QRIO, 5
Questions and answers (Q&A), 22–50
  battery, 22–23, 25
  copy/convert video (movie conversion), 37, 40–45
  incompatible data music files, 33, 36, 38–39
  Memory Stick, 26–30
    (See also Memory Stick)
  movies, 24–26
    (See also Movies)
  music, 23–24
    (See also Music)
  photographs, 30–32
    (See also Photographs)

remote control, 48–50
screen brightness, 23, 37
Wi-Fi, 41, 46–48

Race craft, 51
RadioShack, 201
RAR, 124
RARLAB, 124
RCA phono male-to-male adapter plugs, 201
Reassembling the PSP, 71
References, 215–218
Reign of Fire, 211
Remote control, 48–50
Rengoku: The Tower of Purgatory, 206
Resident Evil: Apocalypse, 87
Right-hand button cluster, 67
Riker, Tim, 186
Ripped DVD movies, 42–43
RnSK Softronics, 122
Robosapien, 95, 96
Rolling your PSP apps, 123–124
ROM image copying, 117
Roshal, Alexander, 124

Sales figures, 5–7, 9, 87
Samurai Camploo, Episodes 1 & 2, 211
SanDisk Image Mate 12-in-1 reader/writer, 28, 29
SanDisk Memory Stick PRO Duo media, 27–29
SanDisk MobileMate for Memory Stock reader/writer, 29
Saw, 213
SCEI, 7
Screen brightness, 23, 37
Scripting language, 123–124
SD/MMC media cards, 173, 174
SECA, 7
SECA Web site, 51
Service set identifier (SSID), 47
Sharp LCD, 57
Skinning the PSP, 126–132, 140–143
Slideshow, 31, 50
Sonny, 2
Sonus, 2
Sony Computer Entertainment, Inc. (SCEI), 7

Sony Computer Entertainment America (SCEA), 8
Sony Corporation:
  black/silver color design, 4
  corporate name, 2, 4
  Morita, 1, 3, 4
  product failures, 5
  SCEA, 8
  SCEI, 7
  Totsuko, 1–4
Sony DPP-EX50, 190–193
Sony Ericsson, 73, 76
Sony Ericsson P800, 74
Sony Memory Stock Adaptor, 190, 192
Sony PlayStation game library, 12
Sony UMD and Memory Stick keeper, 133
Sound Button, 23–24
Sound Grip, 138, 139
Spider-Man 2, 87, 206
SSID, 47
Static IP, 52
Stereo 1-watt PSP amplifier, 95–102
Stereo sound, 95–102
Steve Oedekerk Presents: Thumb Wars and Thumbtanic, 213
Stuffit Deluxe, 124
Subpack adapter, 182, 183
Suppliers, 215–218

Target, 162
Template (PlayGear Pocket skin), 142, 143
Terminal, 33, 36
"Test Connection" option, 48
Text editing app, 123
TFT LCD, 57
Theseis: Journey into Hades to Unveil the Mystery, 206, 207
THM, 41
3GP Converter 031, 44
Thumbnail, 41, 45
Tiger Electronics, 186
Tiger images, 136, 141
Tokyo Telecommunications Engineering Corporation, 1
Tokyo Tsushin Kogyo K. K. (Totsuko), 1–4
Tone settings, 24

Top ten games, 203–207
Top ten movies, 209–213
Torche ELWIRE, 108
Totsuko, 1–4
track7games, 206
Tradition (leather), 133
.TRASH file, 33

UMD, 87–104
  available games, 102–104
  available movie titles, 91–94
    (See also Movies)
  basis anatomy, 19–20, 25
  firm upgrades, 25–26, 90
  inserting/ejecting, 26
  multiplayer games, 88
  official approval, 87
  popularity, 87–88
  sales figures, 87
  stereo sound, 95–102
  swap hack, 89–90
  where to find the movies, 88
UMD access door, 89
UMD disk drive cover, 20
UMD games, 88
UMD movies (see Movies)
UMD movies titles, 91–94
UMD release switch, 20
UMD swap hack, 89–90
UNIQUE (sound button), 24
Universal Media Disc (see UMD)
USB-charged batteries, 159–162
USB connectivity freedom, 170
USB-host port, 164
USB on-the-go (OTG), 157–171
  Audiovox PVR1000, 170
  battery charger, 159–162
  digital camera, 159
  iAUDIO X5, 162–171
  jukebox, 170
  photographs, 169
  PSP + PS2 (Metal Gear Acid), 171
  USB storage device, 170
  USB-View utility, 159
  video camcorder, 169
USB pin outs, 158
USB port, 18, 157

USB storage device, 170
USB-View utility, 159
Used PSP, 60

VAIO, 5
Vaja, 133
Vaja i-volution leather suit PSP case, 133–136
Vaja online customization system, 133
Value Pack accessories, 125–126
Video camcorder, 169
Video camera, 169, 173–177
Video conversion, 37, 40–45
Video games, 113–124
  availability—U.S. PSP launch, 14
  emulating other game systems, 119–122
  GBA media player (GBAmp), 115–118
  health risk, 197–198
  historical overview, 113–115
  homebrew games, 12–13, 118–124
  milestone products, 114–115
  multiplayer games, 88
  portable playability, 12
  PSP + PS2, 171
  PSPE, 124
  RAR, 124
  rolling your PSP apps, 123–124
  sales figures, 5–7, 9
  testing the game, 124
  top ten games, 203–207
  UMD games, 88
  users, 6
Video playback, 175
VideoNow Color, 186–187
Volume, 25

Wallen, Brandy, 136
Wardriving system, 146
Warranty, 58, 60
Web addresses, 215–218
Web browser, 51–55
WEP, 47, 48
WEP key, 48

Where to find it (suppliers, games, products), 215–218
White, Chris (Whitey), 127
Wi-Fi, 41, 46–48
  (See also Wireless network)
Wi-Fi compliant, 41
Wi-Fi hotspot, 147
Wi-Fi Seeker, 147
Wi-Fi sniffer, 146, 147
Wi-Fi snooper, 147
Wi-Fi switch, 18, 52
Wicked Dragons, 136
Wild card, 36
WinRAR, 124
Wipeout Pure Web hack, 51–55
Wired equivalent privacy (WEP), 47, 48
Wireless-B Game Adapter, 147, 148
Wireless Ethernet bridge, 147
Wireless network, 41, 46–48, 145–156
  Junxion Box, 151
  PS2, 147, 148–149
  PS3, 196
  PSP, 152–156
  Wi-Fi sniffer, 146, 147
  wireless Ethernet bridge, 147
  wireless router, 147, 150
  WLAN, 46, 47
Wireless router, 147, 150
WLAN, 46, 47
WLAN switch, 46
WowWee Ltd. Robosapien, 95, 96
Wrist strap, 125
www.elinux.org, 186
www.vajachoice.com, 133
www.wipeoutpure.com, 53

X-OOM Fit-to-Stick technology, 37
X-OOM Movies on PSP, 37, 40, 42
X-ROM driver, 117
XMB, 23
XrossMediaBar, 23

Ys: The Ark of Naphishtim, 206

Are you on the go?

Stop worrying about your camera's memory, just plug your camera into iAUDIO X5 and store all of your photos!
Play your music, video and pictures anytime, anywhere!

Enjoy various features of iAUDIO X5 such as Video/Audio Player, FM Radio, Voice Recorder and Picture Viewer.

- 1.8-inch 260,000 **Color LCD**, Built-in HDD (20GB/30GB/60GB) [1]
- Compatible with Most Digital Cameras through the **USB OTG** [2]
- Play up to 14 hours (35 hours for iAUDIO X5L) [3]
- Supports **MPEG4 Video** up to 15 FPS [4]
- **Picture Viewer** [5]
- **USB 2.0** High-speed Download (480Mbps)
- Supports **Multiple Audio Codecs** including OGG/MP3/WMA/FLAC
- Supports **Portable Hard Disk**
- High-quality **Voice Recording**, Listen to and Record **FM Radio**
- Superb sound quality with BBE/Mach3Bass/MP Enhance/**5 Band Equalizer**/3D Surround

● **Portable Multimedia Player**
iAUDIO X5
Size* / Weight **
4.08 X 2.39 X 0.56 inches / 5.11oz (X5)
4.08 X 2.39 X 0.72 inches / 6.35 oz (X5L/X5 60GB)
* excluding the LCD   ** including the built-in battery

Save 10%
Save 10% with any iAUDIO X5 purchase
Visit www.cowonamerica.com/daveprochnow

Toll Free 1-888-453-8283   Email : tech_support@cowonamerica.com
www.cowonamerica.com/daveprochnow

1. 1GB means 1 billion bytes. Actual formatted capacity less. Playable media files : 2,000 folders and 10,000 files. 2. Some models of digital camera or reader may not be supported. Visit www.cowonamerica.com for the list of cameras supported. 3. Based on company's self-test, The playing time may be shortened according to settings. 4. With JetAudio VX (Video Conversion Utility) 5. Supports JPEG. Progressive JPEG is not supported.

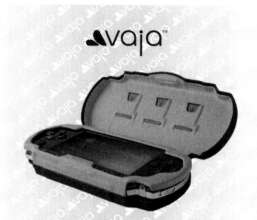

MAIL-IN REBATE COUPON * SAVE UP TO $20

Purchase any Vaja product at **vajacases.com**
or **vajachoice.com** and receive a $20 rebate

Name: _____
Full Address: _____
_____
_____
Country: _____
Telephone: _____
E-mail: _____
Order Nº: _____
Order date: _____

Mail coupon to:
Vaja Corp.
"Mail-in Rebate"
3058 NW, 72 Ave.
Miami, 33122, FL
USA

Conditions: This coupon is only applicable to orders processed until June 1st, 2006. Maximum allowable refund is $20 per coupon. Only one coupon may be used per purchase order (minimum amount: $120). The rebate will be applied to the credit card account used for the purchase shortly after the reception of this coupon. Offer expires 06/01/06. Offer available for all products exhibited in vajacases.com and/or vajachoice.com. This offer may not be combined with any other current offer and/or promotion from Vaja. This coupon may not be duplicated or reproduced in any way. Fraudulent submission could result in Federal prosecution under U.S. mail fraud statutes. Employees of Vaja Corp. (or family or household members), affiliated companies, licensed retailers and distributors are not eligible for this offer. Privacy policies of vajacases.com and vajachoice.com apply on this offer.

gamer graffix™

GET SKINNED!

GG50J — PROMOTIONAL CODE

Lifetime Warranty — GEEK CHIC — "Anything Else is JUST a Sticker!"™

1 FREE JEWEL SKIN

WWW.GAMERGRAFFIX.COM

LOG ON TO WWW.GAMERGRAFFIX.COM. CHOOSE YOUR SKIN.
AFTER PLACING PAYMENT, TYPE PROMOTION CODE IN THE MESSAGE BOX.
THE HOLDER OF THIS COUPON IS ENTITLED TO 1 FREE JEWEL SKIN